The Military Frontier

TIME
LIFE ®
BOOKS

Other Publications:
WEIGHT WATCHERS® SMART CHOICE RECIPE COLLECTION
TRUE CRIME
THE AMERICAN INDIANS
THE ART OF WOODWORKING
LOST CIVILIZATIONS
ECHOES OF GLORY
THE NEW FACE OF WAR
HOW THINGS WORK
WINGS OF WAR
CREATIVE EVERYDAY COOKING
COLLECTOR'S LIBRARY OF THE UNKNOWN
CLASSICS OF WORLD WAR II
TIME-LIFE LIBRARY OF CURIOUS AND UNUSUAL FACTS
AMERICAN COUNTRY
VOYAGE THROUGH THE UNIVERSE
THE THIRD REICH
THE TIME-LIFE GARDENER'S GUIDE
MYSTERIES OF THE UNKNOWN
TIME FRAME
FIX IT YOURSELF
FITNESS, HEALTH & NUTRITION
SUCCESSFUL PARENTING
HEALTHY HOME COOKING
LIBRARY OF NATIONS
THE ENCHANTED WORLD
THE KODAK LIBRARY OF CREATIVE PHOTOGRAPHY
GREAT MEALS IN MINUTES
THE CIVIL WAR
PLANET EARTH
COLLECTOR'S LIBRARY OF THE CIVIL WAR
THE EPIC OF FLIGHT
THE GOOD COOK
WORLD WAR II
HOME REPAIR AND IMPROVEMENT
THE OLD WEST

This volume is one of a series that examines
various aspects of computer technology
and the role computers play in modern life.

UNDERSTANDING COMPUTERS

The Military Frontier

BY THE EDITORS OF TIME-LIFE BOOKS

TIME-LIFE BOOKS, ALEXANDRIA, VIRGINIA

Contents

The World of War Games

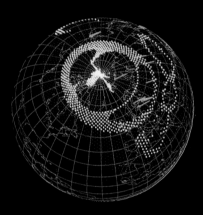

Computers have revolutionized virtually every aspect of modern warfare, from surveillance and weapons systems to communications and battle management. But their impact is nowhere more apparent than in the proxy world of war games—combat simulations used for the very serious business of honing skills and analyzing tactics.

War games exist for all manner of military operations, from small-scale battles to global engagements, with computers contributing to the play in a variety of ways. During field exercises, for example, they act as sophisticated scorekeepers, carefully tracking the live action and calculating the effects of dummy weapons fire. In most cases, however, their role is to create a kind of electronic game board for simulated encounters, in which moves consist of keystrokes and the action unfolds on a computer screen.

Computerized war games are constructed from mathematical models that employ numbers and equations to represent the myriad factors of a combat situation, such as the terrain, weather conditions, and the characteristics of vehicles, weapons, and fighting units. Algorithms, or problem-solving procedures, then manipulate these numerical descriptions to determine the outcome of individual moves or an entire game. Powerful computers are essential. To run the model of a futuristic space-based defense system shown here, for example, a computer had to calculate a stupendously complex set of geometries describing the trajectories of ballistic missiles and the positions of a swarm of satellites designed to destroy them in flight.

Testing a satellite shield. The computer-graphics image at right shows one proposed arrangement for a satellite defense system *(yellow dots)* for the United States, and the predicted path of Soviet missiles from launch to target *(orange lines)*. In the three scenes above, the larger dots indicate satellites that are in position to fire at the particular instant shown; the pattern changes as the missiles follow their course toward North America, allowing analysts to gauge the effectiveness of this particular deployment scheme.

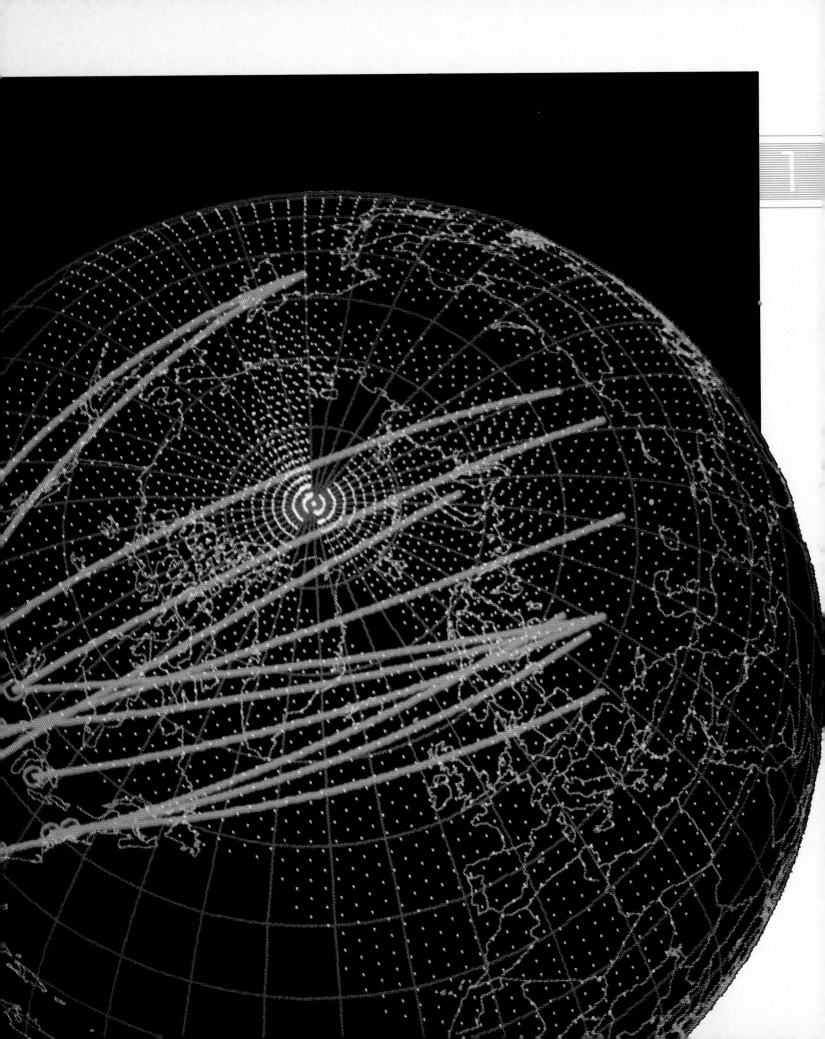

The scenario. Superimposed on a computer image is Blue's two-pronged attack plan (SLEDGE and HAMMER) on objectives DOGFISH and BLUEFISH, blocked by a Red minefield *(crosshatches)*. Vertical phase lines (PL) are references for synchronizing movements through the battle area, between the horizontal boundaries.

Engaging the enemy. As the divided Blue force begins its advance, an infantry unit is hit by enemy artillery *(red line with rectangle)*. White slashes across vehicles indicate that their positions have not been updated in the last fifteen minutes because their signals have been blocked by dust or terrain.

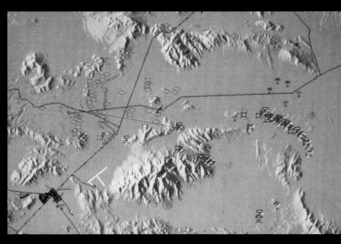

Capturing the Details of Combat

The U.S. Army's National Training Center (NTC) at Fort Irwin, California, serves as a permanent home for a time-honored form of war-gaming—mock battles involving real soldiers and military equipment maneuvering over real terrain. A computerized monitoring system now gives observers a high-tech vantage point on such exercises; via computer-generated images, they get a detailed view of the action as it happens and can also replay it later for further study.

Army units from throughout the United States travel to the NTC to vie against a force trained in the use of Soviet arms and tactics. (The Soviet stand-ins are designated Red, the U.S. forces Blue—a convention often used in war games.) During an exercise, vehicles and selected soldiers representing infantry units carry transmitters that send identifying signals to small receiving stations scattered over the gaming area. The signals are forwarded to a central computer, which calculates positions by triangulation and then indicates them on a computer monitor as symbols: squares for tanks, diamonds for ar-

mored personnel carriers, single and double crosses for artillery, umbrellas for radar, and stick figures for infantry units.

Computers also take care of simulating weapons fire, primarily through the Multiple Integrated Laser Engagement System, or MILES. Weapons such as tanks fire harmless laser beams at their targets, and players and vehicles equipped with MILES receivers register hits by detecting the beams; each beam is actually a series of light pulses coded by weapon type, so that a shot from an M16 rifle, for example, will not be interpreted by a tank's MILES receiver as a kill. Artillery shots, whose arcing trajectories cannot be simulated by lasers, are calculated by the computer from range and bearing tables. Both shots and hits are transmitted to the central computer for recording and display.

Computerized charting of field maneuvers helps reveal flaws in an attack or a defense. Here, Blue's inability to coordinate an advance through the mountains—clearly illustrated by the computer images—causes its assault to fail.

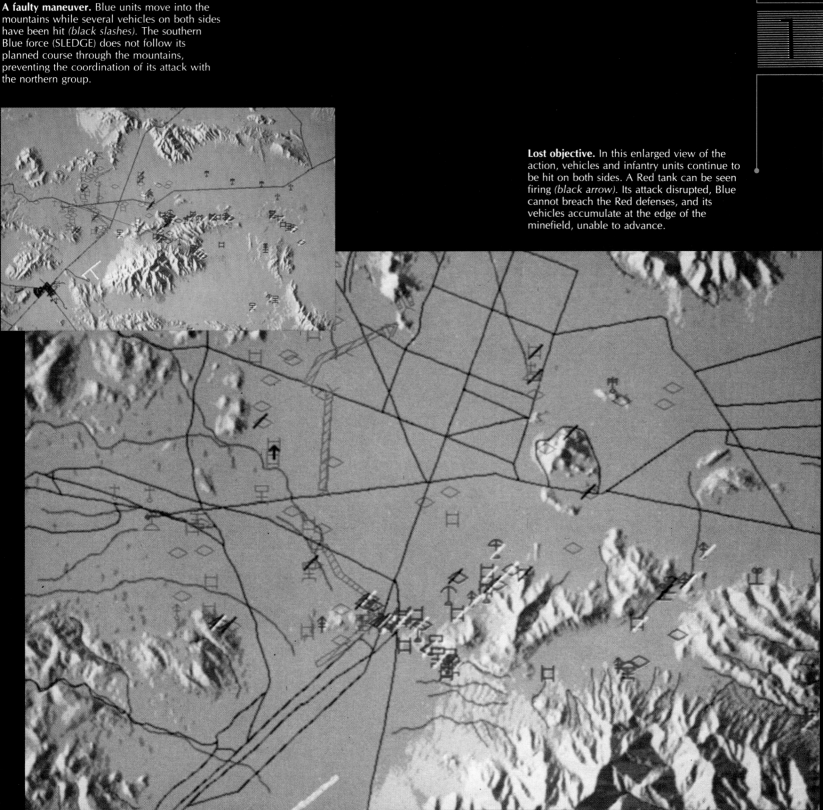

A faulty maneuver. Blue units move into the mountains while several vehicles on both sides have been hit *(black slashes)*. The southern Blue force (SLEDGE) does not follow its planned course through the mountains, preventing the coordination of its attack with the northern group.

Lost objective. In this enlarged view of the action, vehicles and infantry units continue to be hit on both sides. A Red tank can be seen firing *(black arrow)*. Its attack disrupted, Blue cannot breach the Red defenses, and its vehicles accumulate at the edge of the minefield, unable to advance.

Jamming an Attempted Crossing

Intercepting a message. Various graphics symbols depict a simulated Red advance on a Blue force holding high ground *(green contour lines)* across a river; orange lines represent roads. A Red artillery unit radios a message *(yellow lines)* to other units, the text of which appears at the bottom of the screen. One of Blue's direction finders (DF) detects the transmission *(white line)*.

Striking with artillery. With the location of the transmitting Red artillery unit pinpointed, a Blue artillery unit near the river *(blue box with white dot)* can target and destroy it *(blue line with burst)*. The Blue artillery head-quarters (HQ) records a longer-range strike on another Red artillery unit.

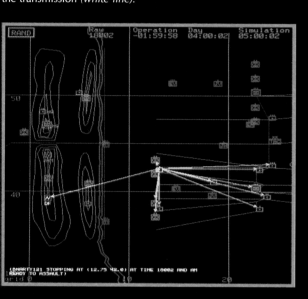

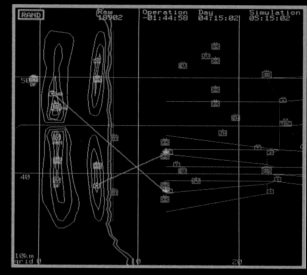

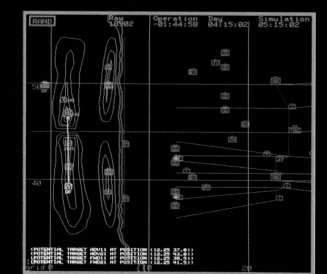

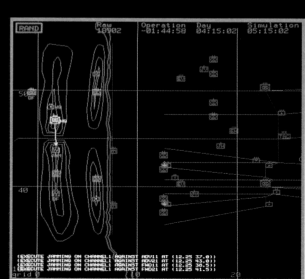

Simulating Electronic Warfare

Communications during battle have always been critical to combat performance. A computerized war game called TWIRL, for Tactical Warfare in the ROSS Language, re-creates through pure simulation the sophisticated electronic measures and countermeasures that are available to today's battlefield commander.

ROSS is known as an object-oriented language because it models the behavior of objects individually and sets up interactions between them. Each piece of military equipment being simulated exists as a separate program chunk that exchanges data with other object programs to represent exchanges that might actually occur in combat. ROSS is thus ideally suited for an emphasis on communications; objects transmit data to other objects just as one field unit might send messages to another.

The playing pieces in TWIRL include direction finders and jamming units for detecting and blocking enemy transmissions. Gamers take the role of commander and experiment with different mixtures of electronic and physically destructive warfare to find the most effective approach. Here, Blue's combination of artillery fire and jamming successfully disrupts Red's attempt to cross the river.

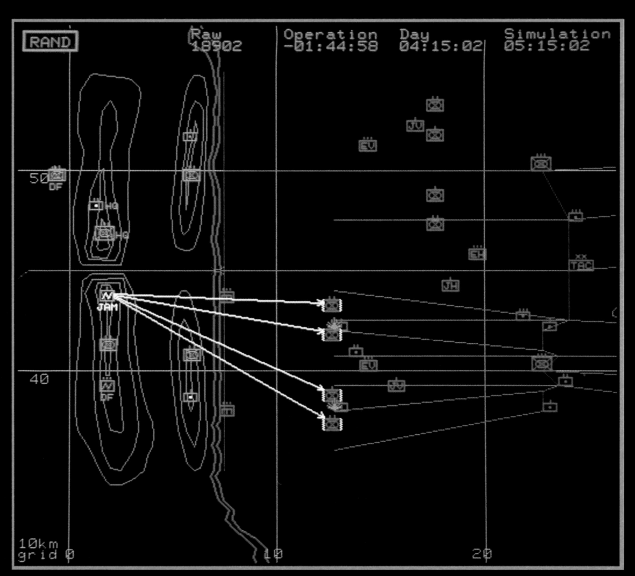

The advance stalls. The Blue unit begins jamming *(white lines)* its specified targets; white brackets signify that the jamming tactic has succeeded. The rifle units are now unable to communicate among themselves or with other Red units, sowing confusion in a pivotal forward rank of Red's attack.

A Two-Sided Test for Battalion Commanders

Janus is considered to be a classic among computerized war games. Named for a Roman god whose two faces pointed in opposite directions, it offers two different views of battalion-level engagements, in which each side's perspective is limited as it would be on an actual battlefield. An extensive data base and the ability to handle millions of programming instructions per second enable Janus computers to incorporate a wide array of factors into their simulations, further adding to the realism of the game.

The players or teams representing opposing commanders

Blue and Red Views of Battle

The yet unseen enemy. The displays at right show separately the Blue and Red forces positioned for combat on the simulated battlefield—a contoured landscape complete with buildings, towns and roads *(yellow clusters and lines)*, woods *(green)*, open areas *(black)*, and a river *(blue line)*. The right side of each screen is a menu from which players select moves with a cursor *(white cross)*. The Blue display *(near right)* reveals Blue units but no Red opponents because there are currently no line-of-sight contacts between the combatants. The Blue commander positions his cursor over a village to mark it as the focus of a subsequent detailed view *(below)*. The Red commander's view of the area *(far right)* shows his forces distributed on both sides of the river. He has selected "Show" with the cursor, asking for a display of the planned advances *(dotted orange lines)* of several units.

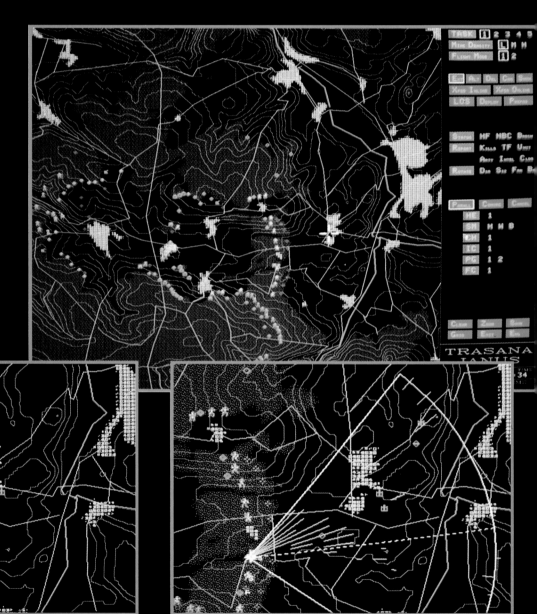

Spotting the enemy. In a close-up view centered on the cursor, Blue has sighted a Red helicopter and tank near the village. Red artillery fire *(orange bursts)* lands near Blue defensive strongpoints *(purple crosses)*, and some units are suppressed (S), or temporarily immobilized. Symbols .... Blue Task Force 1.

Checking a viewpoint. The Blue commander examines one unit's field of view, a wedge-shaped area divided into two sectors by a dotted line. Uneven terrain and buildings obstruct lines of sight *(orange rays)* in the top sector, and only a Red troop carrier *(diamond)* is spotted. Other Red units displayed .... sighted by other Blue forces.

initially see only their own forces on their separate computer displays. Each side controls hundreds of individual units—tanks, helicopters, infantry groups, and the like—which it maneuvers over the computer's simulated landscape. The computer plays out the commanders' moves by processing information from its data base about the operating characteristics and the combat performance of these units, as well as the effect that features of the terrain will have on both movement and field of view; the computer then alters each side's display accordingly. As the game proceeds, the opposing forces eventually come within sight of each other, and the battle is joined.

Janus' power is apparent in the range of options it provides its players through relatively simple on-screen controls. As the illustrations below indicate, each commander plots maneuvers and directs attacks, keeping abreast of developments by ordering close-ups of different battle sectors or even a particular unit's current field of view. The computer fulfills all requests, adjusting its displays in a flash and always ensuring that each side sees neither more nor less than it should.

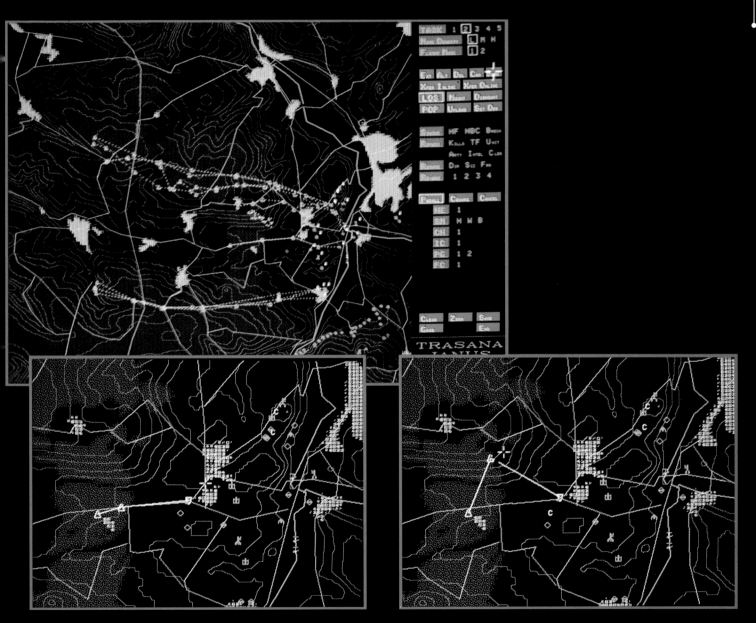

Outlining a flight path. The Red commander has used the cursor and menu controls to plan a helicopter's route, inserting white triangles into the display; the computer draws the connecting lines. The first triangle is inverted, which tells the computer not to begin the flight yet. In the north, two Red units are labeled as casualties (C).

An altered course. The Red commander moves the second triangle of the helicopter's route and repositions the cursor for a closer look there, at the same time keeping the flight on hold by not righting the first triangle. The Red troop carrier that was sighted earlier by Blue (opposite, bottom right) has since been destroyed (C).

A Game to Assess a Real-World Scenario

Few places of strategic importance to the United States are as remote as Iran, a nation rich in the oil so critical to the economies of America, Europe, and Japan. A politically unstable nation, Iran could become the stage for conflict involving the United States. With such a possibility in mind, military planners continually evolve rapid-deployment strategies to get U.S. troops into the region soon enough to be effective. The war game TACWAR from the Institute for Defense Analyses provides the analysts with a ready means of assessing their plans.

Unlike TWIRL and Janus, which create generic landscapes as settings, TACWAR accurately models real locations. The game divides a region such as Iran into smaller pieces called battle areas. Each battle area is associated with data that includes the locations of major cities and other fixed sites such as air bases, bridges, tunnels, and command centers, mountains, and other natural or man-made barriers to move-

ment. As TACWAR simulates combat between opposing forces, battle areas also provide the backdrop for troop locations, movements, and engagements.

When a player changes the disposition of forces, or their assignments, TACWAR processes huge amounts of detailed information to determine how the new circumstances affect each side's rate of advance and the likely outcome of combat. The computer makes these computations so quickly that many days of action can be simulated in minutes.

Results of actual TACWAR games are classified, so the sequence illustrated here is a demonstration based on a subtly unrealistic scenario. In this case, TACWAR tests whether Blue forces deployed on a particular schedule to certain locations would be enough to prevent Red forces from capturing Iranian oil fields along the Persian Gulf. The results suggest that, after more than 80 days of combat, this particular use of Blue forces is sufficient to halt the offensive short of the oil fields.

A Forceful Attack from the North

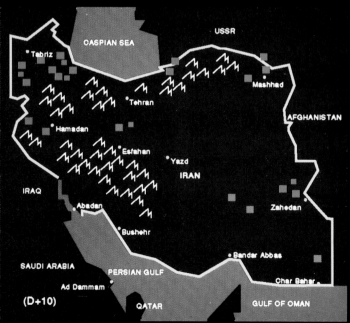

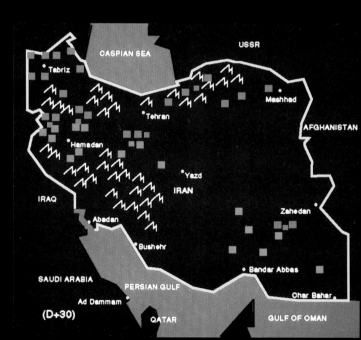

The invasion begins. The TACWAR map pinpoints Red and Blue forces ten days into the invasion. Squares of different sizes show the location and relative strength of each unit. Red forces following a pre-existing mobilization plan continue to arrive. Blue commanders react by deploying various quick-reaction forces into the region.

Pressing the attack. By day 30, Red forces have taken the northern mountains, while in the south, they head for the port city of Bandar Abbas. Mountainous and desert terrains limit the number of troops that can use a major route of advance. Meanwhile, more Blue forces arrive and move to engage the enemy.

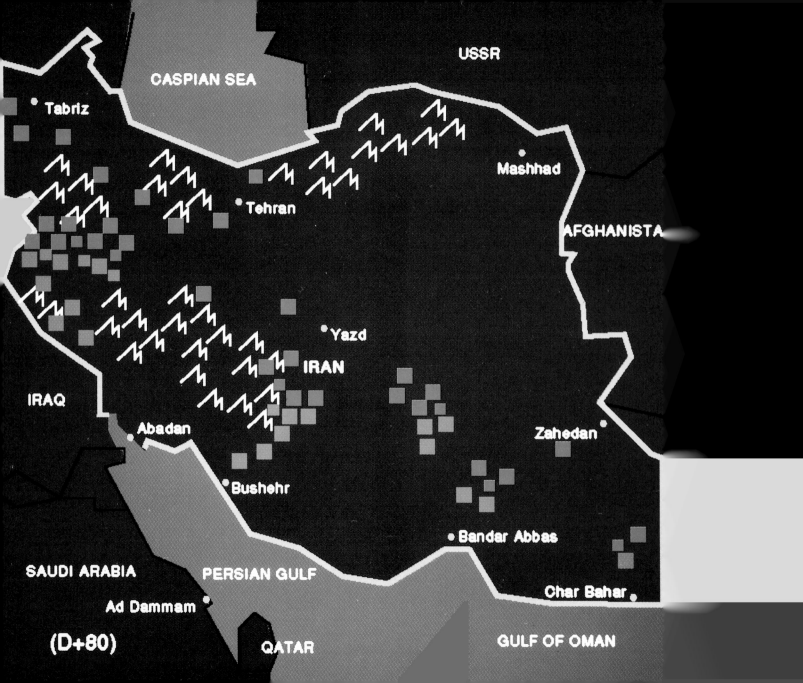

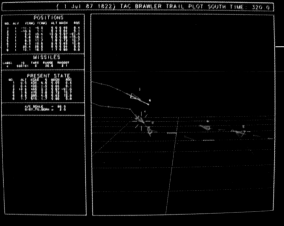

Flights of Fancy

War-game computers model not only ground-based scenarios but aerial ones as well. Some games help train pilots by letting them "fly" simulated aircraft. The two games shown here, however, are designed for analysts, providing a medium for evaluating air-war tactics. As usual, the computer's main role is to represent reality as accurately as it can, but its involvement in play is more extensive than in many other games; in essence, the computer takes over the game, automating all moves based on preprogrammed instructions from the player.

To simulate aerial combat (left), TAC (Tactical Air Combat) Brawler manipulates computer models of supersonic fighters, factoring in their aerodynamics and control systems, as well as typical pilot responses. The player contributes by adjusting variables before the simulation begins—changing, for example, the weapons or fuel load, the number of fighters involved, or the aircraft's detection capabilities. Running the model, the analyst can then assess, say, the effect of plane weight on performance or the significance of situation awareness.

SWIRL, for Strategic Warfare in the ROSS Language (page 11), examines the difficulties U.S. bombers might encounter in attempting to penetrate Soviet air-defense systems (right). Again, the player only takes part before the simulation runs, programming strategies and then watching as the computer plays out the game.

In addition to generating statistical analyses, both games present a wealth of information in graphic form. In TAC Brawler, plane locations, speed, and status, along with missile information, are listed to the left of the graphics display, with its red and blue aircraft, yellow missiles, and green rectangles to indicate the ground. The images portray flight paths and aircraft attitudes during climbs, turns, and other combat maneuvers. The SWIRL scenes at right include both airborne and ground-based radar, fighter bases, surface-to-air missile sites, filter centers for relaying information, and command centers.

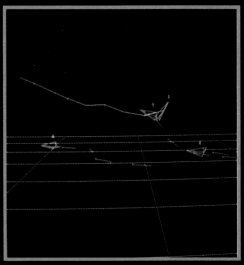

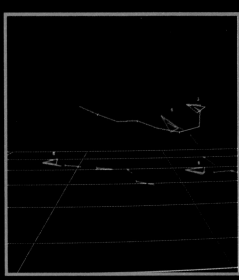

A TAC Brawler Dogfight

At top, two of Blue's eight aircraft (list, left) descend on three Red fighters. One Blue plane fires a missile (yellow line), while the other is destroyed. Having missed with its first shot, the Blue craft begins to maneuver for another attack (middle), eventually reaching an advantageous position above and behind its Red target (bottom).

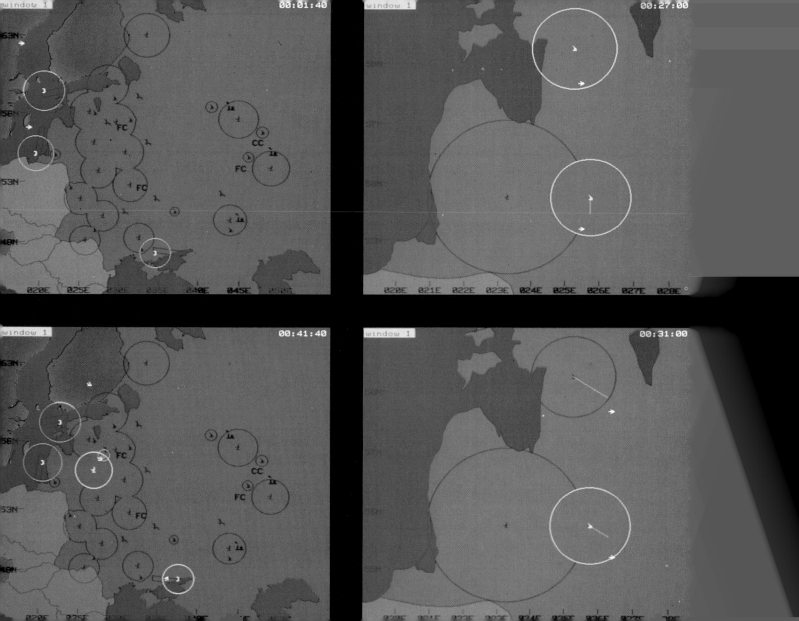

A Computer Version of Warfare at Sea

The U.S. Navy has developed a powerful and flexible computer tool to help analyze operational problems and train its decision makers. The Enhanced Naval Warfare Gaming System (ENWGS) supports a wide range of gaming options, employing intricate models of actual vessels and real-world environments to simulate all kinds of encounters at sea, from local engagements to worldwide naval conflict.

ENWGS can vary a number of factors besides the size and location of battles. To test the skills of the players, for example, the computer may be programmed to incorporate into the model unpredictable sea conditions or erratic vessel performance, mimicking the haphazardness of the real world. When the emphasis is on evaluating tactical procedures, however, randomness is downplayed so that the same moves will produce similar results every time. In addition, participants either face another human team or challenge the ENWGS computer itself, which can effectively simulate the moves of an opposing commander.

As in Janus, ENWGS players see the action from their side only, although a game director monitors a complete view of the game in order to supervise play. The ENWGS sequence on these pages depicts a NATO-Warsaw Pact confrontation involving surface ships, submarines, and airplanes. The pictures portray the engagement as seen by Blue (NATO), Yellow, and the omniscient game director.

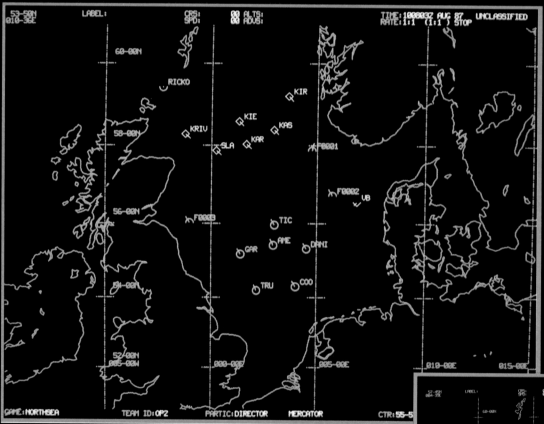

North Sea Encounter

The director's perspective. The display at left, viewed only by the game director, details the positions of all Blue and Yellow game pieces as the action begins. The basic symbols are circles for Blue surface ships and diamonds for Yellow ones, with the upper and lower halves of those shapes representing airplanes and submarines respectively; each unit's current heading is portrayed by a line protruding from the symbol. At game time 803 *(upper right)*, a Blue reconnaissance plane (F0001) unknowingly approaches the Yellow force, while Blue sub RICKO lurks in the rear.

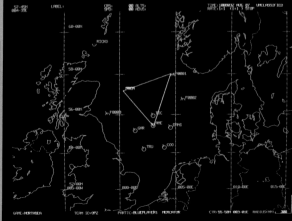

Blue's view of the battle. The Blue commander sees only the disposition of his own forces because none of the Yellow team's ships or submarines has yet been spotted. Blue has requested a display of F0001's flight plan *(white triangle)*, which indicates that the plane originated at and will return to the aircraft carrier AME. During the second leg of its triangular search, it will detect several of Yellow's surface ships and report their positions to the Blue commander.

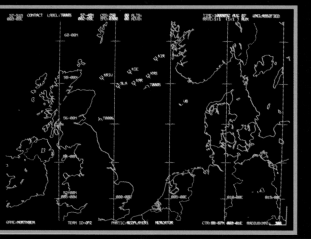

Partial awareness. The Yellow team's display *(left)* shows that only two Blue aircraft have been detected. The computer has turned T0005 (Yellow's designation for Blue F0001) white to signify that Yellow has not yet confirmed its identity as an enemy plane. Messages at the top left of the screen provide additional information on the intruder, such as its latitude, longitude, course heading (CRS), and speed, which will be helpful in targeting a missile attack. Though detected, T0006 (Blue F0003) is too far out of range to be downed by a missile.

Checking range and bearing. With his recon-naissance plane F0001 lost to enemy fire, the Blue commander has instructed the computer to display a compass rose *(magenta web)* originating from Blue ship TIC and covering most of the North Sea. The compass rose will yield range and bearing infor-mation on five of the Yellow surface ships; the sixth ship, to the northwest, and the enemy submarine VB, northeast of the Blue force, remain undetected.

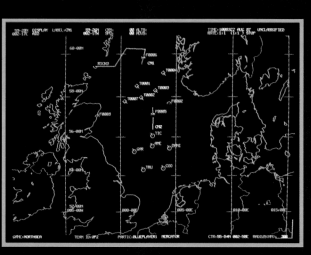

A circuitous course. The Blue commander has launched two guided missiles *(left):* F0005 was fired from the surface ship TIC, and F0006 from the submarine RICKO. They are also labeled CM1 and CM2 to indicate that they are cruise missiles (here carrying conventional warheads). White lines show the missiles' flight paths since launch, and blue lines their current heading. Blue has delib-erately set a roundabout course for missile F0006 to disguise the location of its origin.

Yellow retreat. At game time 901, the director's monitor shows the current status of the overall conflict. As indicated by their change from yellow to red, Yellow surface ships KIE and KAS have taken hits from the Blue cruise missiles, though they are still under power. The Yellow fleet has changed course, unwittingly heading directly toward the Blue submarine RICKO, which has successfully avoided detection.

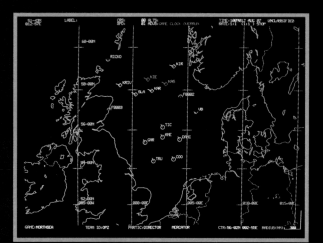

A Fully Automated Rendering of War

The RAND Strategy Assessment System (RSAS) is capable of the ultimate in war games—a computer playing against itself. In this mode, RSAS simulations are completely automatic: All game decisions for both teams are made by the computer, based on complex models of the characteristics and likely responses of the opposing forces. As in TAC Brawler and SWIRL *(pages 16-17)*, humans participate by defining the initial conditions, altering variables in the models in order to assess ways in which different factors might influence the outcome of events. But RSAS plays war on a much grander scale than other games, incorporating land, sea, and air combat into its simulations and depicting action as it unfolds in many diverse regions of conflict.

One of RSAS's most significant features is its ability to take into account political as well as military factors. As a result, analysts can assess the effects not only of alternative military strategies but of different political temperaments. For example, an RSAS Blue model, representing the United States, might be programmed to react boldly to a Red, or Soviet, model exhibiting expansionist tendencies; RSAS would then reveal the most likely consequences, such as whether or how soon nuclear weapons might be brought into play.

The RSAS sequences shown here summarize two different versions of a thirty-day conflict in which Red Warsaw Pact forces invade Norway. The three-frame example on this page assumed a three-week Red preparation, with Blue NATO forces having a two-week warning of the attack. The three pictures at bottom right show the same invasion as a surprise attack. Differences in the results provide valuable information on the penalties associated with being taken by surprise.

A Prepared Defense

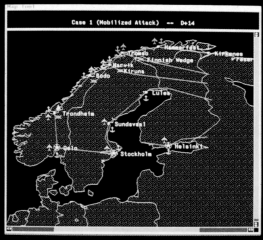

Attack from the north. Of five possible avenues of attack against Norway *(green lines)*, Red has chosen the northernmost two; purple lines indicate forward battle areas. Bridge symbols denote key points in land battles, all of which are currently held by Blue. Two airports *(planes)* and seaports *(anchors)* have been captured by Red.

Action near Tromso. Two weeks into the invasion (D+14), Red ground forces have secured only one key land objective, east of Tromso *(red bridge symbol)*. Red amphibious and airborne forces that earlier seized Tromso meet heavy resistance to the south *(purple line and bridge symbol)*.

A stalled advance. After thirty days of fighting, some Red troops have reached the outskirts of Narvik, as indicated by the purple battle line. Blue amphibious units have re-taken the Tromso port and are in a position to threaten the rear of Red's advance southward.

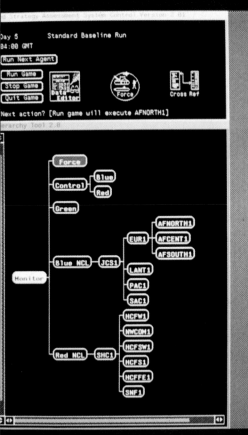

4 - Distribution Maps
5 - Political Cooperation
6 - Strategic Summary Charts
7 - History System
8 - Utilities
->0
Really quit? n

MAIN OPTIONS:
0 - Exit
1 - Regions of Conflict
2 - Central Europe Theater
3 - Far East Theater
4 - Distribution Maps
5 - Political Cooperation
6 - Strategic Summary Charts
7 - History System
8 - Utilities
->0
Really quit? n

MAIN OPTIONS:
0 - Exit
1 - Regions of Conflict
2 - Central Europe Theater
3 - Far East Theater
4 - Distribution Maps
5 - Political Cooperation
6 - Strategic Summary Charts
7 - History System
8 - Utilities
->1

REGIONS OF CONFLICT:
<r> - Quit
1 - Read Force data

Regions of Conflict
(04:00 GMT, Day 5)

NO CONFLICT   CONVENTIONAL   UNMODELED
CONVENTIONAL AIR STRIKES   NUCLEAR

Master control. Analysts watching RSAS simulations control certain features of the computer's operation through the master control panel shown at left. The menu in the upper right portion prompts the user to select displays. Here, a view of the situation worldwide has been called up *(lower right)*; the color-coded map shows regions of conflict in yellow. The chart at lower left indicates that the user has selected the "Force" command from the system's "Monitor" function, which means that the computer will create a detailed record of the status of all military forces throughout the entire game. At upper left are the controls that run the game.

The Element of Surprise

A minimum of resistance. Five days into its surprise attack against Norway, Red ground forces advance toward Hammerfest and Tromso *(purple lines)*. Red airborne and amphibious units have captured airfield and port facilities at Tromso, Narvik, and Bodo.

The push southward. At the two-week point, Red ground forces have reached Tromso and threaten Narvik. Blue has managed to retake the airport near Bodo, but Red's position is still superior along the northern coastline.

An established position. By the end of the month, Red has consolidated its forces at Narvik, seizing key objectives just south of the city and to the east at Kiruna *(red bridge symbols)*. A battle is under way for the seaport at Bodo *(purple anchor)*, but Red is clearly in control to the north.

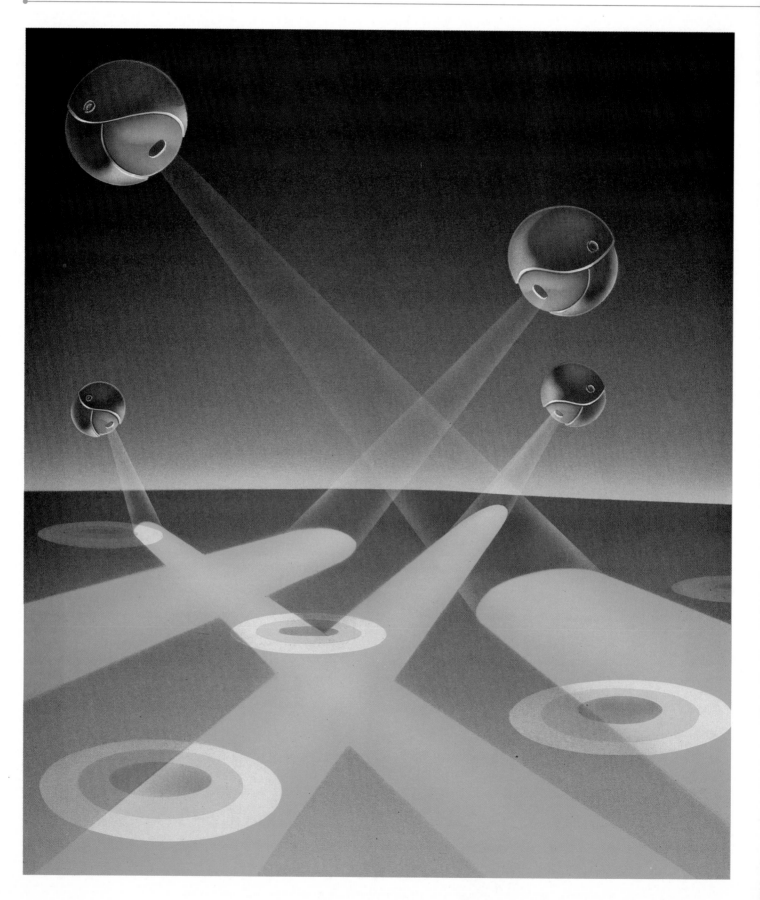

Electronic Sentinels

In the War Room of the North American Air Defense Command (NORAD) headquarters in Colorado Springs, Colorado, the alert came with chilling swiftness: "Coca Color *actual!*"—code words for a nuclear attack on the United States and Canada. The time was 3:15 p.m.; the day, October 5, 1960. Moments before, computers linked to the radar net of the Ballistic Missile Early Warning System (BMEWS) in Thule, Greenland, had announced a flight of missiles rising out of the east. Now there were dozens of missiles, with more appearing every second. The alarm-level indicator in the War Room began flashing "3"—the signal to alert Washington, Ottawa, and the headquarters of the Strategic Air Command (SAC) in Omaha, Nebraska. It quickly went to "4," then "5," meaning a 99.9 percent certainty of attack.

But something did not make sense. No preparations for an attack had been observed at any Soviet military base, and the BMEWS computers had begun to report more missiles in flight than the entire Soviet arsenal contained.

Both the BMEWS radars and the computers that interpreted the signals they received had been in operation only four days. Perhaps one of the two had developed glitches—a diagnosis that the continuing flow of discordant information tended to confirm. More and more objects were filling the sky, the system reported, but they seemed to be barely moving. Not only that, but each radar burst was taking seventy-five seconds to return an echo, when only an eighth of a second would have been required to detect a missile at the 2,200-mile range that BMEWS had calculated. Convinced that BMEWS—not the Soviet Union—was misbehaving, NORAD's commanding general canceled the alert just minutes after it began.

As a later investigation revealed, however, BMEWS had not malfunctioned at all; the system was simply more powerful than anyone had imagined. Thought to have a range of only 3,000 miles, the BMEWS radars had actually reached out nearly a quarter of a million miles; the "missiles" the system had detected were in fact radar returns from the face of the rising moon. Unfortunately, no one had programmed the BMEWS computers to discount the passage of heavenly bodies. And in the absence of any instructions for handling numbers higher than 3,000, the computers had simply divided 3,000 into the distance to the moon and reported the remainder—2,200 miles—as the range to the unidentified object.

By the time the incident was made public two months later, the BMEWS computers had been reprogrammed to ignore lunar echoes, and Pentagon spokesmen were insisting that the United States had come "nowhere near" the brink of accidental war; only the president of the United States, they said, was authorized to launch nuclear missiles on his own. But the story retains its nightmarish quality, especially since it was not the last time a computer failure has fooled the system. At 1:26 a.m. on June 3, 1980, for example, SAC went to worldwide alert when it received an incorrect warning of impending attack by

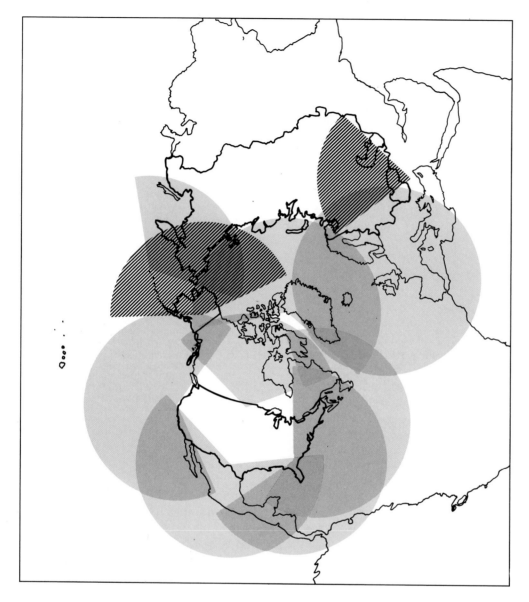

Polar views of the northern hemisphere show the coverage that American early-warning radar *(left)* and the Russian equivalent *(right)* will have in the 1990s. Blue sectors are served by phased-array radars *(pages 26-30)*; cross-hatched sectors represent older types of radar. The range indicated for each installation is the distance at which it can detect missiles or satellites that are more than 500 miles high. U.S. radars facing north watch for Russian ICBMs; radars on the Pacific, Atlantic, and Gulf coasts are alert to attack by sea. Except for outposts in Clear, Alaska, and Diyarbakir, Turkey, the United States plans to replace older radars with modern phased-array installations.

two submarine-launched ballistic missiles. The false alarm was later traced to a malfunctioning computer chip, worth just forty-six cents.

A MILITARY-COMPUTATIONAL COMPLEX

Disquieting as these events are, they dramatize the extent to which computers have been woven into the fabric of the modern military. Computers have transformed the ancient art of war into a twentieth-century science, and neither offense nor defense would be possible today without them.

Vigilance, for example, is no longer a simple matter of posting a keen-eyed sentry atop the nearest hill. In the modern world, threats can come from under the sea, from over the horizon, from missile silos half a world away, and even from space. To cope with such threats, military planners have devised and refined a whole new set of senses—including radar, sonar, and satellite imaging—that let them see the previously invisible.

So efficient are these senses at gathering information, however, that they can flood their users with more data than the human mind can handle. To control

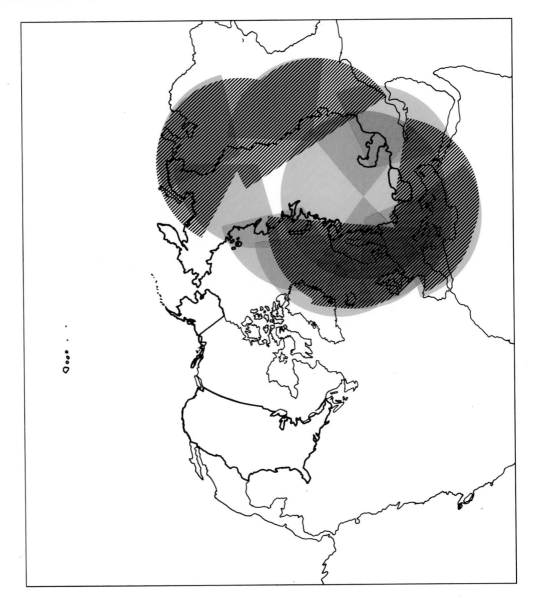

Soviet early-warning radar, which will not extend to the North American continent, is intended to detect an attack launched over the pole by ballistic missiles, as well as an assault from Europe or Asia. At some locations, the Soviet Union plans to supplement—rather than to replace—older radars (cross-hatching) with the phased-array variety (green).

that torrent of information, to sift the meaningful signal from its surrounding noise, and to help human analysts pinpoint crucial facts in time to do something about them demand computer power in huge amounts. It is no accident that the Central Intelligence Agency, the Pentagon, and the National Security Agency have quietly taken the lead in spurring the development of computer technology in the United States.

Meanwhile, the battlefield itself is no longer an arena where individual warriors survive by their wits and reflexes. Events in modern combat occur faster than humans can react to them, and the computer is the cause. The armies of the world have capitalized on the microprocessor to deploy all manner of "smart" weapons—bombs, missiles, mines, and artillery shells—that are guided by their own internal brains of silicon. Some military scientists hazard that warfare may one day be completely automated—a sort of savagely realistic computer game fought with robots and weapons that require no direct human intervention.

Nor is military command any longer the domain of generals surveying the field of battle from a distant jeep or command post. Warfare has become global in scope. Today the commanders of a battle may exercise control from thousands of miles away, using computers to determine the most effective disposition of forces and then flashing orders to the battlefield in the form of digital data. Indeed, in an era when intercontinental ballistic missiles (ICBMs) can reduce reaction times to thirty minutes or less, the automated decision-making capabilities offered by computers have proved an unending source of both temptation and trouble for the military. With each stage of computerization that the armed services reach, computer scientists and military planners will be challenged to ensure that the sort of reasoned judgment displayed that October day at NORAD never becomes obsolete.

A 105-foot-tall phased-array radar installation at Otis Air Force Base on Cape Cod, code-named PAVE PAWS *(left),* provides early warning of submarine-launched missiles. Computer-steerable radar signals *(pages 28-29)* come from nearly 1,800 sprinkler-shaped emitters studded in rows on each of the radar's two faces *(right),* which cover a 240-degree sector of the sky. PAVE PAWS has a maximum range of 3,000 miles; at 1,200 miles, it can detect, identify, and track a target no bigger than a basketball.

ELECTRONIC SENTINELS

NORAD is a good place to begin an examination of computer use in military surveillance. NORAD's mission is to coordinate the defense of North America against attack from above—a category of threats that includes manned bombers, cruise missiles, intercontinental ballistic missiles launched from the Soviet heartland, and ballistic missiles launched from submarines lurking just offshore. To accomplish that mission, NORAD monitors the skies with a series of radar lines that ring North America like so many electronic picket fences. And to keep track of the data that pours in from these listening posts, NORAD operates an imposing array of eighty-seven computers.

As recently as the mid-1980s, however, military professionals were questioning whether NORAD's computers—victims of more than a decade of neglect and miscalculation—could perform their mission. NORAD's computer problems stemmed from an efficiency measure adopted by the Department of Defense in 1970. The department decreed that the Pentagon would make a bulk buy of thirty-five Honeywell 6000-series mainframe computers for its worldwide communications web. Against the advice of NORAD commander General Seth McKee, who objected that the machine was not sophisticated enough, NORAD was assigned a Honeywell 6000 as its main tracking computer. In the words of one congressman, it was "a shotgun marriage ordered by the Joint Chiefs of Staff."

The difficulty lay with the Honeywell 6000's particular brand of processing information. Like most computer makers of the 1960s, Honeywell had designed its machine as a batch processor—a computer that processed data one batch at a time, completing each job in sequence before moving on to the next. But this was precisely the wrong approach for NORAD, whose radars track some 7,000

targets—civilian airliners, orbiting and decaying satellites, miscellaneous space debris, even the moon. (The *A* in NORAD's name now stands for Aerospace rather than Air.) NORAD's computers must constantly identify these harmless targets as a way of distinguishing them from real threats. Otherwise, a commercial airline flight might be misinterpreted as a cruise missile, or a reentering chunk of space junk might be mistaken for an incoming warhead.

The Honeywell batch processor fell far short of keeping pace with NORAD's radars. What NORAD required was a computer system that could monitor rapidly changing events in real time—that is, as they occurred—and offer immediate access to that data. What NORAD received instead from the Pentagon in 1974 was a second Honeywell 6000.

While NORAD cobbled together a variety of support computers to buttress the 6000s, NORAD programmers devised a new operating system that accelerated processing time. Still, the effort was a bit like turning a bus into a sports car, and the results were every bit as predictable. A 1980 report by the Air Force Inspector General branded the NORAD system "an unsuccessful acquisition" that delivered "marginal performance."

In response, a number of fixes were undertaken for NORAD. One of the first beneficiaries was NORAD's Space Defense Operations Center, charged with tracking those thousands of objects in space. The Pentagon set aside $360 million for the upgrade, with a particular focus on providing better data processing, new displays, and improved communications. The result was a computer system known as SPADOC I, which the Ford Aerospace Corporation began assembling in 1982; since then, the SPADOC system has undergone three major overhauls.

Radar for Modern Times

The principle of radar, one of the most important military inventions of World War II, is well known: An antenna sends out a short pulse of radio waves, then listens for a reflection of the signal from a distant object. A conventional radar antenna used for surveillance rotates continuously, ten revolutions per minute or so, to search in every direction. Tracking radar of the conventional type turns to face a single target identified by a surveillance radar and locks onto the quarry in order to direct a missile or gunfire at it.

Although these sorts of radar are adequate for defending against aircraft, they would be overwhelmed by an onslaught of ballistic-missile nuclear warheads numbering in the hun-

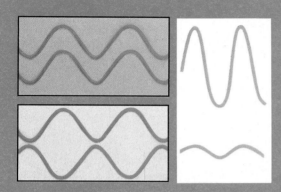

A question of phase. When waves travel peak to peak, that is, in phase with each other, they add up to a wave that is stronger than any individual *(top)*. When waves are out of phase *(bottom)*, peaks and troughs offset each other and create a wave weaker than its components.

Scanning the skies. A phased-array radar detects only objects that fall within a narrow beam where radar transmissions are in phase. At right, an incoming missile warhead happens to be in an area where the radar signals are out of phase, momentarily eluding detection. But as the computer rapidly adjusts the timing of the emitters so that the beam scans downward, the radar illuminates the warhead *(center)*. The computer displays the intruder on a monitor and retains the warhead's position and heading in memory. With that information, the radar can be aimed frequently at the threat to update its position, even as the search continues for other targets *(far right)*.

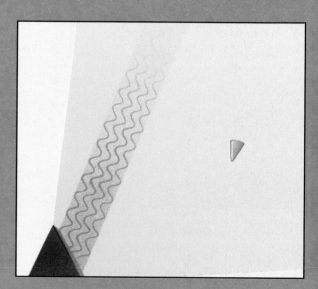

According to Nicholas L. Johnson, an advisory scientist at Teledyne Brown Engineering in Colorado Springs and an expert on NORAD's computers, the current version, SPADOC IV, "processes 70,000 observations a day, every day—and does it quite well."

RADARS MARCHING IN PHASE

Even as NORAD struggled to bring its outmoded computers up to date, the escalating pace of new missile technology threatened to render its radars obsolete. When the three BMEWS radar stations that report to NORAD were erected in Greenland, England, and Alaska in the early 1960s, ICBMs carried only one warhead apiece, and a "major missile attack" constituted fewer than two dozen missiles. Starting in the 1970s, however, both the United States and the Soviet Union began deploying a new generation of ICBMs capable of carrying not only multiple warheads but also an accompanying swarm of decoys. If an attack ever came, the BMEWS radars would be faced with more than 100,000 projectiles—far too many for the system to track reliably. The Pentagon therefore embarked upon a massive upgrading of its ground-based radar system in the mid-1980s.

dreds and traveling at 15,000 miles an hour. That stupendous tracking challenge gave rise to a radically different type of radar in the 1960s. Called phased-array radar, it has no moving parts, yet it can direct defenses against hundreds of fast-moving targets simultaneously, even as it scans the skies for additional intruders.

In place of a single antenna, such a system has row upon row of closely spaced emitters *(page 27)*. The emitters send out signals of equal strength over a broad area of coverage. Most of the waves, being out of phase, weaken each other. But some of them, being in phase, reinforce each other to create a narrow beam of strong radar waves. The beam is steered electronically by the simple tactic of delaying some waves by a few microseconds—a job supervised by a computer. The system then listens for echoes only from the area of the sky where the strong beam has been aimed, ignøring any reflections that might occur when out-of-phase waves encounter an object.

Miniaturization of fast, powerful computers has broadened the application of phased-array radar. Not only is it used on land for early warning of a missile attack, but now it is also installed on ships *(pages 96-97)* and even in aircraft, offering instant-by-instant tracking of almost any combination of attacking forces.

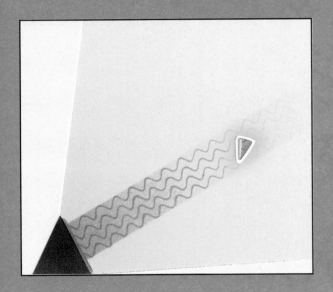

The essential piece of technology in this effort is known as phased-array radar, which many observers see as the most crucial advance in the field since the invention of radar itself. Old-style radar antennae—the rotating dishes once common atop airport control towers—were too slow and unreliable to follow a swarm of warheads. Phased-array radar compensates for that sluggishness by eliminating the mechanically steered antenna altogether. The new system is entirely electronic, employing thousands of tiny antennae spaced evenly, or arrayed, over the face of a motionless wall.

Phased-array radar derives its success from the fact that the bigger a radar antenna is, the narrower (or more tightly focused) its beam can be. Because more energy is focused on (and thus reflected from) the target, a narrow beam permits easier detection of targets. Being small, each individual antenna in a phased-array radar emits a wide beam. But when the antennae radiate together, the resulting merged beam displays the narrow width that would be produced by a single large antenna the size of the entire array *(above)*.

Just as a searchlight covers less space than a streetlamp, however, a narrow beam cannot match the coverage of a wide one; a radar emitting a narrow beam must therefore make do by rapidly scanning with its beam. Such scanning, or steering of the radar beam, is performed by a computer that staggers the times at which each antenna emits its separate signal. When the computer instructs all the radar antennae to send their signals simultaneously, the beam is propagated perpendicular to the array. But when the computer introduces a time delay in the emission of signals from one antenna to the next, the beam leaves the face of the array at an angle. The greater the time delay from antenna to antenna, the more the beam is deflected. The net effect is a narrow, highly steerable beam.

Most phased-array radars are designed to provide early warning of incoming aircraft, ballistic missiles, and cruise missiles. Because of the power output required for this mission, they tend to be huge. The four hulking PAVE PAWS radars—the acronym stands for Perimeter-Acquisition Vehicle Entry/Phased-Array Warning System—erected at air force bases along the coasts of the United States in 1979 are good examples. Each PAVE PAWS radar has two faces; these scan in different directions and operate independently. With each face containing 1,792 antennae and measuring 105 feet in height, the installation is nearly the size of a ten-story building.

Yet it is a marvel of reliability. Because a phased-array radar has no moving parts, mechanical breakdowns are infrequent. And even when a few of its antennae do fail, so many others remain in operation that the radar's overall performance is not undermined. A PAVE PAWS radar is also very accurate and very fast: It can observe targets at a distance of 3,000 miles, and each face can scan its beam across a 120-degree arc of the sky in just a few millionths of a second. This scanning speed makes a phased array uniquely suited to tracking multiple targets, since the computer that steers the radar beam can update its information many times per second simply by flicking the beam from one target to the next before any one of them has moved very far. Still, the PAVE PAWS radars have an important shortcoming: They are no better than previous radars at seeing past the curvature of the earth. Their signals travel in perfectly straight lines, meaning they cannot spot low-flying aircraft or cruise missiles until those targets have flown to within short range.

To deal with this deficiency, the air force in the late 1980s introduced a new system, called over-the-horizon (OTH) radar, that was designed to complement the coverage of PAVE PAWS. Over-the-horizon radars manage to look beyond the curve of the earth by allowing their signals to ricochet between the surface of the planet and the ionosphere, a layer of electrically charged atmospheric particles about 200 miles high. As a result, OTH radars can detect low-flying objects in a band ranging from 500 miles out to about 1,800 miles out. Unfortunately, the over-the-horizon system cannot determine how high the objects are, but its computer can analyze the reflected signals to reveal how fast and on what course the objects are proceeding.

HIDE-AND-SEEK BENEATH THE SEA
In much the same way that NORAD patrols the skies with radar, the U.S. Navy patrols the ocean depths with sonar—an apparatus that detects the presence and location of submerged objects by the sound waves they reflect or produce. Although experts continue to debate whether sonar constitutes an art or a

science, few dispute the fact that it is a pursuit in which computers have come to play a vital role.

It is small wonder that the navy's primary target in this effort is the Soviet Union's submarine fleet. If war ever came, a single enemy attack submarine equipped with homing torpedoes could savage a convoy of surface ships, or a single missile submarine equipped with nuclear warheads could decimate a continent. U.S. Navy vessels have thus been armed with a variety of lethal means for destroying a hostile sub. Attack submarines carry sound-homing Mark 48 torpedoes; destroyers carry sound-homing Mark 46 torpedoes of their own, as well as antisubmarine rockets, called ASROCs, that release torpedoes or nuclear depth charges upon entering the water; and carrier-launched aircraft carry nuclear depth charges. First, however, such hunters must find their quarry.

Any commander trying to track a submarine with sonar faces a tactical decision: active or passive? Active sonar works very much like radar. The vessel emits a high-pitched pulse of sound—a "ping"—that travels outward through the surrounding water. When the sound encounters an obstacle—be it a submarine, a whale, or a rock—some of it is reflected back to the vessel, where it is analyzed by onboard computers. The direction of the returning sound waves indicates the direction to the target, while the time that elapses between the emission of the pulse and the return of its echo reveals the distance to the target. The strength of the reflection may also tell the sonar operator something of the size—and even the shape—of the target vessel.

Yet active sonar has disadvantages, especially when one submarine is hunting another. It has a relatively short range; and in a contest where stealth is everything, the pings betray the hunter's position. Active sonar, if it is used at all, thus tends to be reserved for the end game of a submarine hunt, when the hunters are moving in for the kill. Most of the preliminary tracking is done by passive sonar—a high-technology version of staying very quiet and listening.

Where active sonar is conspicuous, passive sonar is cumbersome. To use it effectively, a surface ship or submarine must proceed at a trawler's pace, often trailing a long cable bristling with sensitive underwater microphones called hydrophones. From a single point in the sea, most passive sonar systems can determine only the direction, not the range, to a target. In order to find that distance, the listening vessel must move to at least one other point, again determine the direction to the target vessel, and triangulate the target sub's position from those measurements.

Passive sonar, however, has the major advantage of being covert: With no ping to give away the hunter's position, the quarry is often unaware that anyone is following his progress. A second virtue is that passive sonar can detect noisy targets at great distances. About 1,500 to 4,000 feet deep in the ocean there exists a natural "sound duct" where sound waves become trapped by a combination of water temperature and pressure. Loud sounds in this zone can carry an extraordinarily long way. Throughout the 1960s and 1970s, this phenomenon provided a bonanza of intelligence for the United States Navy. "The first time the Soviets put the pedal to the metal on an Alpha boat in the Barents Sea," said one naval expert, "the noise traveled all the way to Bermuda"—some 4,000 miles. Although the Soviets took pains to quiet their new submarines in the 1980s, notably by furnishing them with more precisely milled propellers, passive sonar remains an invaluable detection device for the navies of both superpowers.

Passive sonar can also reveal the identity of submarines in motion. At high speeds, the noise given off by a submarine is characterized by the hiss of air bubbles spun from whirling propeller blades. At low speeds, the noise is dominated by the rumble of pumps and the whine of turbines. These noises form a unique sonic portrait of a submarine, called its sound signature, that distinguishes one class of boat from another. Within each class, the idiosyncracies of each boat's machinery noises may further serve to identify individual subs almost as reliably as fingerprints differentiate human beings.

AN UNDERWATER LISTENING NET

To monitor submarine movements on a global scale, the U.S. Navy has deployed a network of hydrophones known as SOSUS, or Sound Surveillance System. Begun in the 1950s in response to the appearance of Soviet missile-carrying submarines, SOSUS has expanded from a series of underwater listening posts in American coastal waters to a worldwide network. Along the coast of Norway, for example, a SOSUS line listens for submarines moving south from the Soviet naval base at Polyarny (page 34).

Although the details of SOSUS installations are classified, the system probably consists of hydrophone arrays buried a few feet deep in the sea-bottom mud or suspended at depths most conducive to sound propagation. Each array is about 1,000 yards long and appears in a line at intervals of five to fifteen miles. Signals collected by the hydrophone arrays from passing submarines travel by underwater cable to nearby shore-based stations; from there they are beamed by satellite to navy computers that integrate the SOSUS data with information gathered from satellites and other intelligence sources. The result is a comprehensive picture of submarine activity in areas of the world where it could most threaten the United States and its allies.

SIFTING SIGNAL FROM NOISE

Submarine hunters work with the knowledge that the "silent service" plies its trade in a very noisy venue. The sounds from storms at sea, singing whales, far-off earthquakes, and thousands of surface ships crowd into hydrophones in such profusion that "there are days out there when nobody can hear anything," according to Admiral Bruce DeMars, deputy chief of naval operations for submarine warfare. An additional problem for practitioners of antisubmarine warfare is that submarines are getting quieter all the time. Even one of the Soviets' old diesel subs gives off only about one watt of sonic power—a sound roughly equivalent to the broadcast of a transistor radio—and the stealthiest American nuclear boats emit hundreds of times less than that. To detect a submarine amid the ambient noise of the ocean thus demands much more than sensitive hydrophones; it requires a sophisticated technique known as signal processing, in which a computer analyzes even the most minute variations in the sonar signals those hydrophones receive. Often the computer analysis can reveal the otherwise imperceptible sound signature of an enemy sub.

For many years, the technique of digital signal processing was hampered by the equipment that performed it. In the early 1970s, for example, sonar and its attendant computer apparatus were still being built with hardwired electronics;

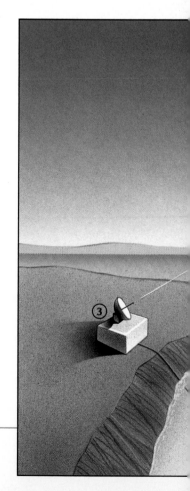

Hunting the hunter. A hostile submarine (1) stalking an aircraft carrier (9) is tracked with a variety of passive and active sonar devices, some fixed and others mobile. At long range, a row of hydrophones on the seabed (2), connected to signal-processing stations on land (3), is used in conjunction with passive sensors towed behind a special-purpose surface ship (4). Floating sonobuoys transmit acoustic information to the aircraft (7) that dropped them.

As the enemy approaches, a patrolling submarine (5) uses onboard computers to process data from passive towed sonar and from both passive and active hull-mounted sonar. A destroyer (8) escorting the aircraft carrier (9) has large, fast computers linked to passive and active sonar. A helicopter (10) lowers its dipping sonar into the water and returns data to the destroyer. Searchers communicate with each other via satellite (11); their findings are sent to a central computer (12) that keeps track of all submarines under surveillance.

Underwater Ears for the Navy

Unlike the surface vessels that are its chief prey, an attack submarine is undetectable by visible light, infrared light, or radar as it prowls the seas hundreds of feet below the surface.

Fortunately for the defense, sound penetrates water to great distances, so the principal systems for detecting submarines, called sonar, are acoustic. Passive sonar listens with devices called hydrophones for telltale noises that even the quietest submarines make. Active sonar sends out pulses of sound, then picks up echoes that rebound from submerged objects.

Passive sonar is able to detect a loud submarine hundreds of miles away and, aided by computers ashore or aboard ships and aircraft in favorable listening conditions, can recognize the boat as friend or foe—and even identify individual submarines by their unique sound signatures. Though passive sonar reveals the direction to a submarine, it cannot readily determine distance. For that information, vital to a successful defense, active sonar is often used. From the direction of an echo and the time it takes to return from a target, a computer rapidly calculates range and bearing. But there is a price: Active sonar divulges its own location in the process.

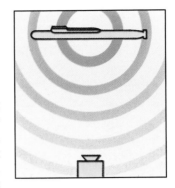

Passive sonar. The noises of submarine machinery and of the hull slipping through the sea, traveling many thousands of times farther through water than they would in air, may well end their journey at sensitive sonar receivers called hydrophones. Once detected, the sounds are first amplified, then isolated from background noise with the aid of computers. Comparing the results with a data base containing patterns of sounds made by various types of submarines is often sufficient to identify the vessel.

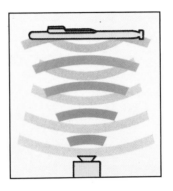

Active sonar. A short pulse of sound emitted by a transducer, a type of underwater loudspeaker, travels through the water and reflects from a target. By measuring the time that elapses between the origin of the pulse and the return of an echo, the computer can determine range with great precision. The direction to the target is determined by aiming the transducer to produce the strongest echo. A target's speed can be calculated by noting the time and distance between successive fixes on its position.

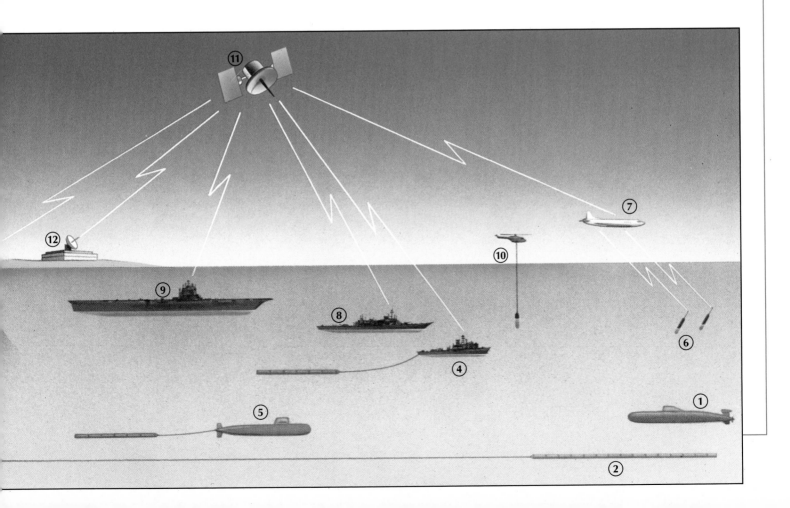

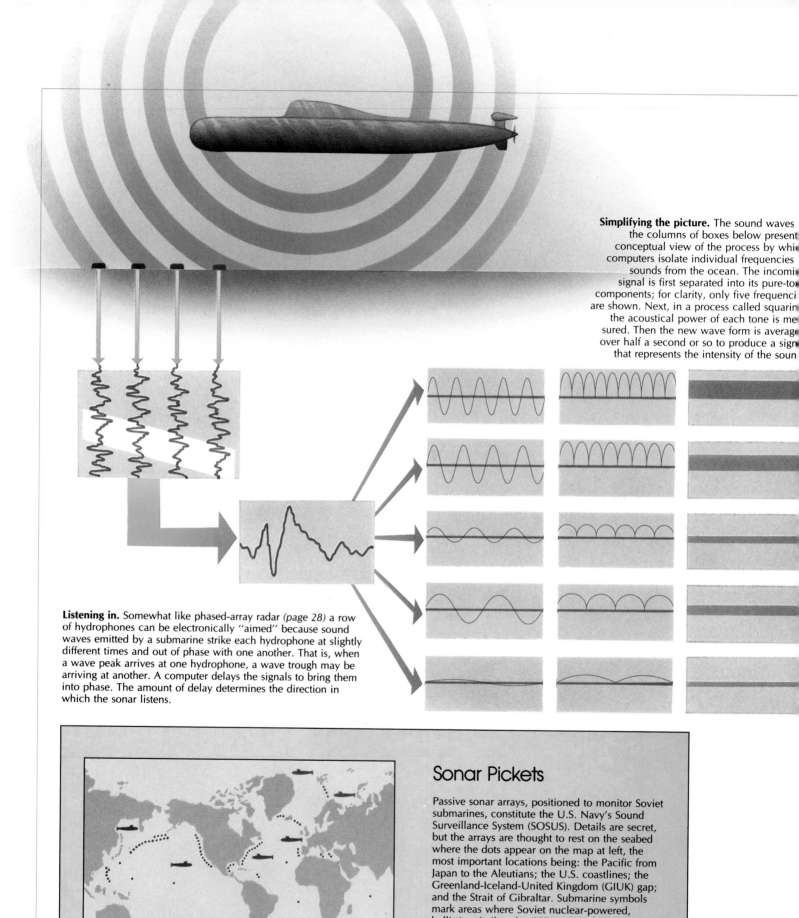

Simplifying the picture. The sound waves the columns of boxes below present conceptual view of the process by whic computers isolate individual frequencies sounds from the ocean. The incomin signal is first separated into its pure-ton components; for clarity, only five frequenci are shown. Next, in a process called squarin the acoustical power of each tone is me sured. Then the new wave form is average over half a second or so to produce a sign that represents the intensity of the soun

Listening in. Somewhat like phased-array radar *(page 28)* a row of hydrophones can be electronically "aimed" because sound waves emitted by a submarine strike each hydrophone at slightly different times and out of phase with one another. That is, when a wave peak arrives at one hydrophone, a wave trough may be arriving at another. A computer delays the signals to bring them into phase. The amount of delay determines the direction in which the sonar listens.

Sonar Pickets

Passive sonar arrays, positioned to monitor Soviet submarines, constitute the U.S. Navy's Sound Surveillance System (SOSUS). Details are secret, but the arrays are thought to rest on the seabed where the dots appear on the map at left, the most important locations being: the Pacific from Japan to the Aleutians; the U.S. coastlines; the Greenland-Iceland-United Kingdom (GIUK) gap; and the Strait of Gibraltar. Submarine symbols mark areas where Soviet nuclear-powered, ballistic-missile subs (SSBNs) patrol. SOSUS may be able to detect a loud submarine up to several hundred miles away and fix its position within a ten-mile circle, but the system may overlook a quiet sub more than a few miles away.

Examining Sound for a Signature

Passive sonar surveillance, such as that performed by a global network used by the U.S. Navy *(box, below),* helps analysts to catalog the sounds of submarines and later identify them wherever they are found.

At high speeds, noises are produced by the sub's propeller; at low speeds, most sounds come from machinery. Some contain a wide spectrum of frequencies; others consist of a few discrete frequencies, or even a single frequency. For example, a nuclear-powered submarine of a specific class might emit a sound like a musical note from its reactor-cooling pump. Unique noises resulting from idiosyncracies such as a slightly unbalanced propeller shaft can be used to identify a specific vessel.

Submarine noises are usually mixed with acoustical clutter from other sources—whales, dolphins, and even surface vessels, for example. To make a submarine stand out from the background noise, signal-processing computers first eliminate most of the noise by directing the sonar to listen in a specific direction. This is accomplished by using a row of hydrophones called an array *(left).*

Next, the computers disassemble the sound picked up by the sonar into its component frequencies. The strength of each is calculated and shown on a special type of monitor known as a waterfall display *(below).* To identify a target, frequency patterns are entered in a computer and compared with a data base of frequencies known to be emitted by various types of submarines.

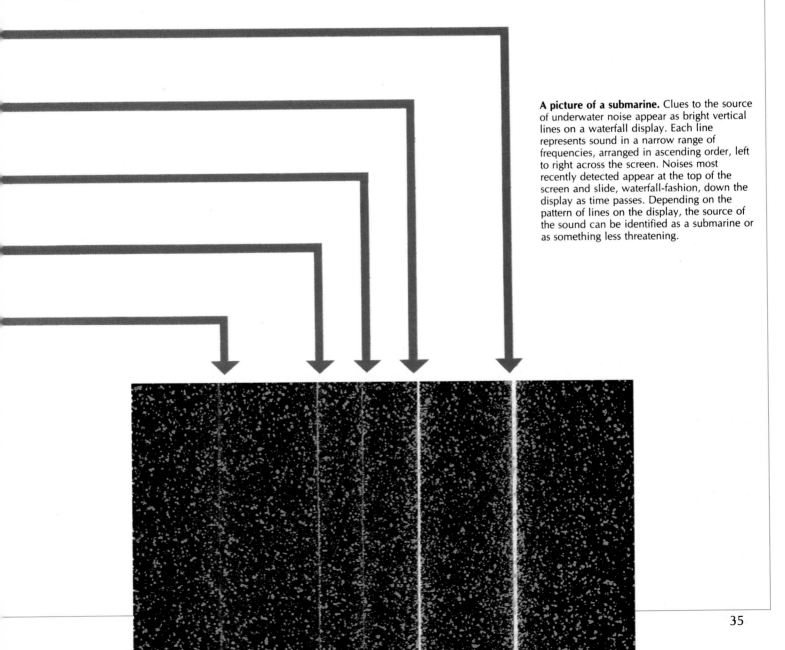

A picture of a submarine. Clues to the source of underwater noise appear as bright vertical lines on a waterfall display. Each line represents sound in a narrow range of frequencies, arranged in ascending order, left to right across the screen. Noises most recently detected appear at the top of the screen and slide, waterfall-fashion, down the display as time passes. Depending on the pattern of lines on the display, the source of the sound can be identified as a submarine or as something less threatening.

the manufacturer designed each piece of electronic gear to perform one specific signal-processing calculation and to work with one specific set of sensors. A dramatic change in the operating environment—the deployment of a new sensor, the introduction of a new signal-processing technique, the christening of a new class of Soviet submarine—could dictate that the electronics be overhauled or replaced. This built-in obsolescence prevailed until the navy substituted powerful, fully reprogrammable computers for hardwired equipment.

For all their computational intensity, signal-processing techniques work best when coupled with the tracking talents of a skilled sonar operator. To increase the chances of detecting a lurking sub, the operator might follow a divide-and-conquer strategy, instructing the computer to slice up the hydrophones' output into very narrow bands of frequency. No one band will contain a disproportionate amount of the ocean's random noise, but one or several of them may contain a great deal of the characteristic noise made by a particular submarine: On the operator's computer display, the frequency bands containing this noise will stand out sharply, helping to define the sound signature—and therefore the identity—of the target sub.

In areas of the ocean that are unusually noisy, however, the computerized approach can founder. Soviet missile submarines may loiter beneath the edge of the polar icecap, where the shifting ice generates grinding noises that can mask a submarine's passage. In this cacophonous realm, a sonar operator trying to distinguish signal from noise must proceed gingerly. He can tell the computer to ignore all sounds softer than a certain threshold, but where should he set that threshold? With the threshold set too low, the sonarman will be plagued by false alarms caused by random bursts of ocean noise that exceed the threshold. With the threshold set too high, the sonarman could fail to spot a sub waiting in ambush. It is a riddle that computers—lacking the sonarman's essential experience, judgment, and skill—have not yet been able to solve.

Experts predict that few benefits will result from further computerization of the sonarman's trade; sonar instruments cannot be made much more sensitive than they are, and beyond a certain point in the refinement of such listening devices, the sea is simply too noisy. Therefore, the U.S. Navy has begun contemplating a return to active sonar. At the same time, it is examining nonacoustic detection methods—looking for localized distortions in the earth's magnetic field caused by a submarine below the surface, or surveying the oceans for the subtle surface ripples that betray a submarine's wake. Both such methods depend on the most computer-intensive instruments of all—satellites.

SPIES IN THE SKY
The scandal that erupted when Francis Gary Powers and his high-flying U-2 reconnaissance plane were shot down over the Soviet Union in May of 1960 curtailed American attempts to photograph the heart of that country from aircraft, yet it hardly blinded U.S. intelligence gatherers. An even more covert means of peering into the Soviets' top-secret military sites existed in the Discoverer series of spy satellites, legacies of a shared vision that the CIA and the air force had clandestinely pursued since March of 1955. Although they were cumbersome to retrieve and, by today's standards, crude in resolution, the first satellite photographs of the Soviet Union—recovered from Discoverer 13 just three

months after the U-2 incident—sparked a demand for more revealing images that computers helped to fulfill.

Long before computers came to play the central role in enhancing and interpreting satellite photographs, however, the mere recovery of those images posed a daunting technical challenge. The satellites in the Discoverer series ejected their film in a canister that descended by parachute through the earth's atmosphere over the Pacific Ocean, where it was snagged in midair by a C-119 airplane trailing a trapeze bar at the end of nylon cables. A less demanding means of recovering images was inaugurated with the launch of the U.S. spy satellite Samos II on January 31, 1961. Photographs taken by Samos II were developed on board the satellite, scanned with a television-like device, and then beamed back to earth by radio. Unfortunately, the resolution of such images—the size of the smallest detail discernible in them—was relatively poor: about twenty feet. The photographs from Discoverer 13, by contrast, displayed a resolution on the order of twelve inches.

U.S. imaging satellites have undergone dramatic evolution since the era of those early sentinels, achieving a high degree of technical virtuosity in the Keyhole 11 type, first launched in 1976. Essentially a massive telescope pointing downward, the KH-11 weighs about fourteen tons and is approximately the size of a railroad boxcar. Under ideal weather and atmospheric conditions, the resolving power of its telescope can be as fine as four inches. That is not quite good enough to read the title of a newspaper held by a Russian in the street, as is sometimes claimed, but it certainly suffices to make out such details as the identification numbers on the wings of a Soviet fighter plane.

Unlike Discoverer 13 and Samos II, the KH-11 carries not a foot of film. It captures images instead by means of a mechanical retina known as a charge-coupled device, or CCD, a chip of silicon containing an array of tiny electronic cells. When exposed to light, the individual cells collect varying amounts of electric charge much as individual buckets covering a field would collect varying amounts of water during a rainstorm. The stronger the light at any point in an image, the greater will be the charge that registers in the cell assigned to measure that point. The pattern of lights and darks in the image therefore translates into an equivalent pattern of high and low electric charges in the array.

To record the image, integrated circuits associated with the CCD array sample each cell in turn and measure its charge. An analog-to-digital converter on-board the satellite then translates each measurement into a string of eight binary digits, called a byte; 00000000 represents black, 11111111 represents white, and other bytes represent all 254 shades of gray in between. The bytes are transmitted to a computer on the ground, which converts each byte into its corresponding brightness level and displays that value as a pixel—short for picture element—at the appropriate point on a video monitor. The result is a copy of the original image.

A newer version of the KH-11 known as the KH-12 carries a fuel supply, enabling the satellite to maneuver into an orbit over any part of the world on short notice. Experts believe the KH-12 can also use the propellant to dip from its customary height of 175 miles to within ninety miles of the Earth, yielding close-up images with twice the resolution of its predecessor.

Although the interpretation of a given image is ultimately the responsibility of

No Place
to Hide

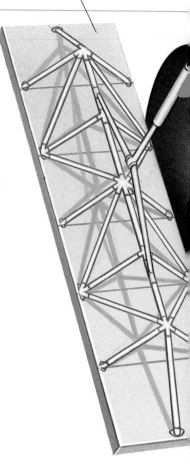

Military reconnaissance satellites, such as the United States' KH-12 illustrated at right, are closely held secrets; no official source will confirm details of their design, operation, or capabilities. Yet there is little doubt that these satellites make use of the latest in computer technology for their surveillance of the world below.

Photographic film, for example, has been replaced in satellites by charge-coupled devices (CCDs), light-sensitive silicon sensors that take digital pictures in amazing detail. Black-and-white shots are made with a single CCD; color pictures are taken with a multispectral system that uses separate CCDs to record blue, green, red, and infrared light. Special radar supplements the CCDs to make clear pictures through cloud cover.

All this data is transmitted in digital form by way of communications satellites to ground-station computers, which reconstruct the overhead view on monitors. Image-processing software improves clarity by correcting distortion in the pictures that results from light passing through the atmosphere and by manipulating color and contrast to make details stand out clearly.

A likely design for the KH-12 reconnaissance satellite employs a high-power reflecting telescope and a variety of electronic sensors to capture images of the earth. An angled mirror can be oriented by an electric motor to permit the telescope to look straight down or to either side. Incoming light is focused by a series of dished mirrors onto a pyramid mirror that diverts the rays to three types of CCDs. One variety is the kind sensitive to visible and infrared light; another senses longer-wavelength infrared rays radiating from hot objects; and the third is an ultrasensitive variety designed to allow picture taking by starlight. The imaging radar system beams radio waves at the earth and creates a picture by examining the echoes. An onboard computer transmits data and receives navigation and telescope-aiming instructions. Solar panels provide electricity to operate the satellite.

This rare view of a Russian shipyard shows the Soviet Union's first nuclear-powered aircraft carrier, the *Kremlin,* under construction in two sections. The picture was made with the CCD imaging system on board an American KH-11 spy satellite, predecessor to the KH-12. Operating at an altitude of approximately 200 miles, the KH-11 can clearly depict objects ten centimeters across, enabling analysts to identify types of armament and radar on the ship. An even more detailed examination is possible with pictures taken by the KH-12. Images from military satellites are never intentionally released to the public; this one was given to *Jane's Defense Weekly,* a British magazine, by a U.S. Naval Intelligence employee.

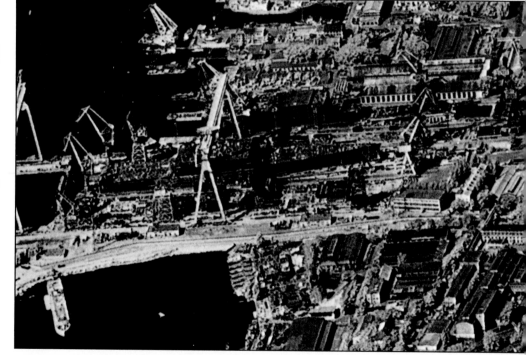

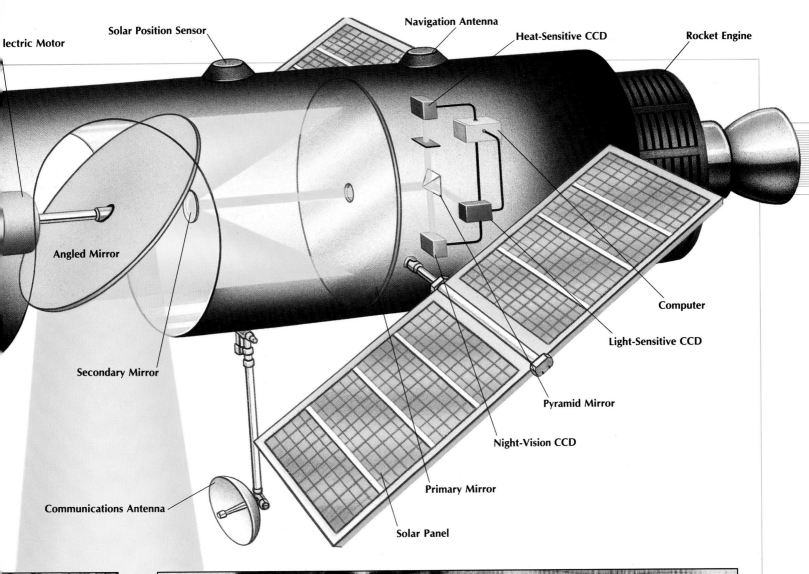

Electric Motor

Solar Position Sensor

Navigation Antenna

Heat-Sensitive CCD

Rocket Engine

Angled Mirror

Secondary Mirror

Communications Antenna

Computer

Light-Sensitive CCD

Pyramid Mirror

Night-Vision CCD

Primary Mirror

Solar Panel

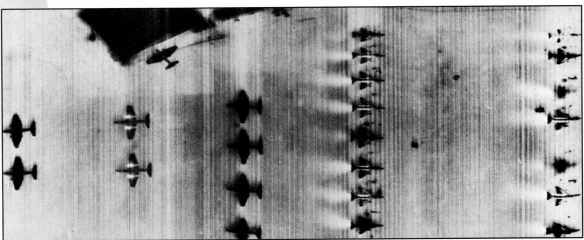

The thermal infrared picture above, taken from an aircraft, shows the degree of detail that military satellites probably see from hundreds of miles above the earth. In a thermal image, warm areas appear light in tone, cool areas dark. A jet with engines running has bright lines between its wings and fuselage. The cool shadow of an aircraft on the tarmac *(far right)* testifies to the plane's recent departure.

Resolving power, the ability of an optical system to produce images of small objects, greatly affects the utility of satellite photographs. The picture at left of a Soviet submarine base on the Kola Peninsula shows little detail; it was made by America's civilian observation satellite, Landsat, a system that muddles details less than thirty meters across. In contrast, a photograph of the same area taken by SPOT, a French commercial satellite having ten-meter resolution, shows roads, buildings, and piers.

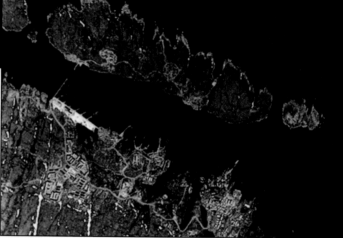

Computers can turn unrevealing photographs into informative ones. A SPOT image of the Krasnoyarsk area in the Soviet Union (above, left), indicates construction suspected of being a phased-array radar, but the fuzzy image is inconclusive. When a computer sharpens the edges of objects, a process that removes color, the radar building—the prominent rectangle at the left of the installation—is clear enough for analysts to confirm its purpose.

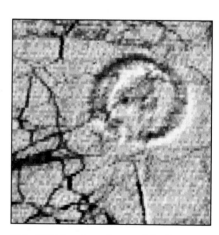

This 1984 Landsat picture of the icy East Siberian Sea, according to Defense Department analysts, shows a circular contrail, two miles in diameter, made by a Russian observation plane. Broken ice at the center of the circle is thought to be the work of a Soviet submarine, testing gear that would allow it to fire missiles from under the polar ice cap.

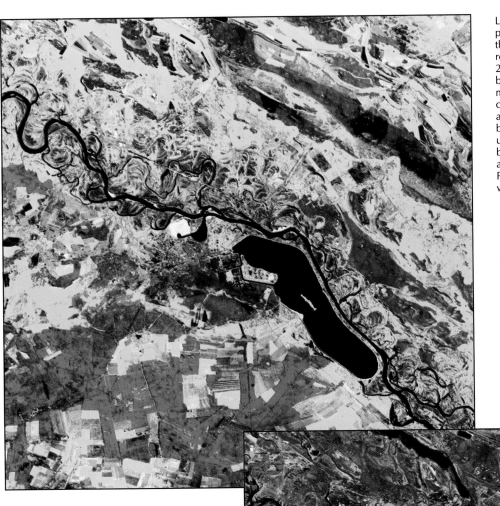

Landsat photographs of the nuclear power plant at Chernobyl, its cooling reservoir, and the surrounding countryside reveal effects of a reactor accident that occurred there on April 26, 1986. In the picture at left, taken a year before the event, arbitrarily assigned colors mark flourishing crops *(bright yellow)* and coniferous forests *(red)*. A year after the accident *(below),* the land lies barren. Dull blue-gray and pale yellow tones indicate unplanted fields. Healthy trees have died or been cut down in cleanup operations. Dark and muted tones along the banks of the Pripyat River show the blight caused by waterborne radiation.

A single stereo pair of pictures—images of a scene captured from slightly different positions—provides sufficient information for a computer to synthesize a three-dimensional view. The stereo pair at left shows an aerospace center on Long Island, New York. From the differences between the two images, a computer program called the Automatic Topographic Mapper can construct any view of the scene. The one below shows the facility as it would appear from a vantage point 300 feet high. Objects smaller than a basketball can be distinguished.

High-resolution radar aboard the space shuttle pierced the clouds that usually shroud Mount Shasta, California, to take these pictures of the summit that were later colored by computer. Software developed at the Jet Propulsion Laboratory in Pasadena, California, was used to combine them into a computer-generated, low-altitude flight around the mountain. One frame of the resulting film appears at right.

Jet Propulsion Laboratory computers mated a Landsat picture of Los Angeles (above) with data about the terrain supplied by the Defense Mapping Agency to create a three-dimensional computer model of the city, its suburbs, nearby mountains, and the San Andreas Fault. The picture at left is part of a film simulating a jet-speed reconnaissance of the model; to produce the whole film required almost six days of processing on a mainframe computer.

professional intelligence analysts, computers can do a great deal to refine the image and ease that interpretation. One reason why satellite photographs must be "cleaned up" by computer is that no optical system is ever perfect. Even with a telescope as sophisticated as those aboard the KH-11 and KH-12, for example, the distortions that occur as an image is reflected from one mirror to the next within the telescope mean that a straight line in the real world may be rendered as a slightly curved line in the image, while a circle may appear as a slightly flattened oval. On the other hand, the extent of those optical effects can be precisely measured when the telescope is built. Once a reconnaissance satellite is functioning, computers on the ground process each telescope image in a way that applies the optical distortions in reverse, straightening out the curved lines and unflattening the ovals.

Even after distortions have been corrected, however, the image may still be almost unusable. Shadows may obscure the key parts of a target, or low contrast may make it difficult to pick out the target from its background. Such problems can often be overcome by enhancing the contrast. To do that, a computer first determines precisely how many pixels the image contains at each of 256 brightness levels. This yields a statistical range of the contrast, which the computer can then analyze by various mathematical processes. For an image with reasonably good contrast, for example, the computer analysis will reveal that a fair number of the points are very bright, a fair number are very dark, and the rest are distributed more or less evenly in between. For an image with low contrast, however, the analysis will show that the majority of brightness values are tightly clustered together. The brightness values of an image made under dim light might cluster around a low number, say sixty-four. In an image taken through high haze, which reflects a great deal of light, the values might be found to congregate near 200.

To redress the imbalance in images that are too dark or too bright overall, the computer shifts the brightness peak to the center of the scale by adding brightness to each pixel or subtracting brightness from each one. Alternatively, the computer might redistribute the tightly clustered peak to fill the available brightness range. A pixel that was only slightly brighter than the average suddenly becomes much brighter, while a pixel that was only slightly darker than average becomes much darker. The result is an image with much greater contrast, which in turn allows for easier interpretation.

FROM CUSTODIAN TO ANALYST

Once the computers have corrected the images and heightened their contrast, human analysts set about interpreting the features the images contain. One way computers can help achieve this goal is to synthesize a three-dimensional perspective from a pair of overlapping, or stereoscopic, overhead views. This is possible because many satellite photographs are taken in sequence, with about a 60 percent overlap from one image to the next. Any feature that falls within this overlapping zone will therefore appear in both images. Computers compare the two images point by point to produce a detailed three-dimensional model of the scene. This model can then be displayed on the analyst's computer monitor and rotated at various angles, coaxing a building to emerge from its shadow or enabling a radar installation to be measured for height.

Another powerful tool for image analysis is color, which often allows an analyst to tell objects apart that might look identical in black and white. It is for this reason that modern surveillance satellites often return images as a series of frames: one for blue light, one for green, one for red, and several for radiation in the infrared region of the spectrum, which is invisible to the human eye.

A computer is crucial for analyzing this rainbow of data. An analyst struggling to divine whether a dark splotch in an image is a patch of vegetation or a camouflage tarp, for example, can instruct the computer to display the green-light frame on his monitor, then superimpose upon it the infrared frame. Because healthy vegetation is highly reflective of infrared wavelengths (the infrared bounces off the sides and corners of leaf cells in the deeper layers of the leaf), the vegetation in the superimposed display normally shows up as bright red, the camouflage appears dark green, and the analyst can tell the two apart.

Infrared images are among the most useful for intelligence work. They can be used to plot the precise boundaries of water (clear, deep water appears black in the infrared frame) or to pick out paths and roads through fields (the vegetation comes out light while the track comes out dark). The same principle applies to other color bands in the spectrum: A road may appear very light at blue wavelengths but very dark at yellow wavelengths. At each of the various wavelengths, different objects and different kinds of terrain all exhibit their own unique patterns of light and dark. Stored in a computer, these patterns can help identify a puzzling item in an image; the computer evaluates the brightness level of the item in each wavelength of the frame, searches out the matching pattern in its library, and makes the identification. Computer professionals call this pattern recognition.

Military image-processing computers at the Pentagon and elsewhere contain millions upon millions of bits of imagery from all over the world, yet identifying the details contained in those images is often not enough. It is more important to discover what has changed since the last time anyone examined the image. Because railroads are used to transport ballistic missiles, for instance, the presence of a railroad spur where none ran before can warrant a closer look.

The computer is a powerful aid in allowing the analyst to quickly detect minute changes in a complex scene. It can be instructed to sift through the latest image of a site pixel by pixel, comparing each one with the corresponding pixels in an earlier image. By subtracting the two brightness values at each pixel, the computer can instantly produce a map of changed values; pixels that stay the same cancel each other out and show up in black on the map, whereas pixels that shift can be assigned a vivid color that will alert the analyst to the change.

THE ARTFUL SCIENCE OF EAVESDROPPING

Another variety of information about the enemy that can best be gleaned by computer is signals intelligence. SIGINT, as it is known in the trade, includes monitoring what people say to one another by phone or by radio, monitoring radar emissions, and monitoring the data that missiles in flight send back to their tracking stations on the ground.

Of all the SIGINT collection sites available to the United States, space is the most valuable: Listening posts in orbit are impervious to the sort of political vicissitudes that robbed the United States of its SIGINT ground station in northern

Iran in 1979. Thus, alongside its sophisticated imaging gear, the KH-11 satellite has an antenna designed to pick up communications in the areas under surveillance. Because the KH-11 is designed to fly low for a satellite (just 200 miles above the earth, on average), it necessarily focuses on a small area. But since the early 1970s, a series of SIGINT satellites code named Rhyolite, Chalet, Magnum, and Vortex have orbited above the equatorial regions of the earth at the geosynchronous altitude of 22,300 miles, meaning they keep pace with the planet's rotation. From that height the satellites can eavesdrop on a huge swath of the land and sea below, including a major portion of the Soviet Union. Each Rhyolite satellite sports a reception antenna more than forty feet across, which allows it to monitor everything from walkie-talkie communications during Soviet troop exercises to the emissions of Soviet radars to the self-diagnoses broadcast by Soviet missiles during test flights. The Rhyolite antenna can intercept more than 11,000 microwave telephone calls at the same time. Finally, a series of satellites code-named Jumpseat follow high, elliptical orbits that leave them suspended over Siberia for as much as eight hours at a time. They reportedly listen for microwave pulses emitted by Soviet anti-ballistic-missile radars.

The National Security Agency—the agency charged with making sense of SIGINT—has the prodigious task of recording nearly every signal that issues from the Soviet Union: the full daily radio broadcast of every radio station in all the Soviet republics, every transmission to every Soviet embassy abroad, every broadcast to Soviet ships at sea, every transmission to Warsaw Pact military units on maneuvers in Eastern Europe, and the radio traffic of the control tower in every major Soviet airport. Intelligence analysts liken such indiscriminately exhaustive collection to the action of "a vacuum cleaner picking up specks of dust from a carpet." Any one of these specks might be meaningless in itself, but when correlated by computer with myriad other data, it may emerge as a valuable nugget of information.

In the cavernous basement of its headquarters at Fort Meade, Maryland, twenty miles northeast of Washington, D.C., the NSA operates nearly eleven acres of computers and computer-related gear. The equipment ranges from conventional magnetic-tape storage systems to immensely fast supercomputers. Using these machines, NSA analysts bring a variety of analytic techniques to bear. One of the most fruitful employs an accelerated version of a basic computer process—matching two values—to winnow kernels of information from the massive harvest of electronic data that the NSA's worldwide facilities reap each day. NSA computers can scan through intercepted texts and cables at about four million characters per second—equivalent to reading this book in less than one sixteenth of a second. When the computers come across certain key words that match those on a master list in their memory—"submarine," for example, or "troop strength"—they flag the passage for scrutiny by human analysts.

TECHNOLOGICAL INCENTIVES

Ever since the U.S. Army used the first electronic digital computer in America, the ENIAC, to calculate ballistics tables in 1946, the U.S. military has spurred the advance of all sorts of computer technology. The Pentagon's Defense Advanced Research Projects Agency (DARPA), founded in 1958 to oversee the Defense Department's high-technology research-and-development programs, has been a dominant funding agency for computer research in the United States. And the impact of the NSA, though that agency continues to operate with a passion for anonymity, has in many ways been just as pervasive.

In the late 1950s and early 1960s, the NSA's support took the form of direct funding. The most lavish benefits accrued to Project Lightning, begun in 1957 by NSA Director Ralph Canine with the goal of increasing the speed of existing computers 1,000-fold within five years. Canine's motivation stemmed in part from a plea by agency cryptanalysts for better ways to sift through the growing mounds of data being collected by the NSA. Project Lightning ultimately became one of the largest government-sponsored computer-research projects in history, involving expenditures of some $25 million over five years.

Unlike other NSA-sponsored programs, Project Lightning was unclassified, with the results of its research being published in the open literature. The idea was to provide handholds for the toddling U.S. computer industry, and that is exactly what happened. Lightning inspired a variety of discoveries in materials properties, high-speed circuitry, and components fabrication. It also prompted a number of fledgling commercial computer firms to start up or to accelerate their own advanced research. Indeed, many computer professionals point to Project Lightning as one of the fundamental stimulants for the commercial computer boom of the 1960s and 1970s.

At the project's close, the NSA withdrew from direct support of computer development, stating that private industry was now better equipped to propel the state of the art. In its role as a customer with deep pockets, however, the NSA continued to underwrite the computer industry. In 1976 the agency took delivery of the first supercomputer produced by Cray Research of Minneapolis, a company that has since dominated the supercomputer business. Today, the NSA still typically purchases the first production model of new supercomputers from Cray and other manufacturers.

Yet the NSA's hands-off policy on computer development eventually brought it up against a paradox. As more and more supercomputers have been applied to civilian uses such as weather forecasting and oil exploration, the machines' new owners have expressed growing concern that their multimillion-dollar investments in software not be rendered obsolete by some new generation of supercomputer hardware. As a result, the companies that make supercomputers have become exceedingly cautious about how they introduce new technology; they have, in the NSA's view, become "risk averse."

In response, the NSA established a Supercomputer Research Center at its Fort Meade headquarters in November 1984. With a budget of more than $20 million per year and a staff of nearly 100 computer engineers, the center has devoted its energies to a brand of computation known as parallel processing, which promises to outpace existing supercomputers by a factor of 10 to 100.

In a conventional computer, calculations are funneled one at a time through a single processor. In a parallel machine, however, each calculation is divided into dozens, hundreds, or even thousands of pieces, and each piece is then routed to its own processor. These "dedicated processors," as they are called, work on their separate pieces simultaneously, then combine the results to yield the answer in shorter time.

For some problems, such as predicting the flow of air over a plane wing or locating a tank in a leafy forest, parallel computers have already demonstrated dazzling speed. But for other tasks—calculating the trajectory of a missile, for example—it is not always clear just how to segment the problem so that the various processors can work on it. In some cases, in fact, parallel machines have proved to be less effective than their serial counterparts. Does that mean parallel computers will have to be redesigned, or hardwired, for each successive problem? Or does the solution lie in devising more flexible software? This and other programming challenges occupy 60 to 70 percent of the engineers at the NSA's supercomputer center.

PEACE: THE FINAL FRONTIER

No doubt the computerization of surveillance will continue to evolve. For example, a microminiaturization technology known as Very High Speed Integrated Circuitry (VHSIC) now enables some photographic spy satellites to carry onboard computers that perform the bulk of image processing before the data is beamed to the ground. In this way, the Pentagon can obtain refined intelligence almost as soon as the satellite captures the raw images. The latest generation of imaging spy satellites, the KH-12s, probably has this capability. For the long term, surveillance specialists foresee the development of self-reliant reconnaissance satellites that could choose their own targets should their ground-based command stations be knocked out in a war, or that could compensate for certain kinds of damage to themselves. On the ground, meanwhile, artificial-intelligence techniques are being developed that would permit the automatic identification of objects in satellite photographs.

The fact that military computers have enabled each superpower to constantly monitor the other may constitute a powerful steadying influence in the world. Computerized surveillance has made it almost impossible for one side to mount a surprise attack on the other. Too many people and too much matériel would have to be moved around in preparation, too many messages would have to be sent; the other side would be sure to get wind of it. For much the same reason, intelligence professionals doubt that one side could ever secretly violate an arms-control agreement, at least in any major way.

Until the day when computers serve exclusively as tools for peace, however, the military will continue to explore the frontiers of their use as tools for war. Indeed, by the late 1980s that exploration was well advanced, and computer scientists and weapons designers had come up with a name for the deadly result of their joint efforts: the electronic arsenal.

Duels of Deception and Detection

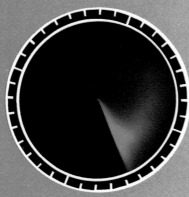

Electronic warfare, which Winston Churchill once called the "wizard war," is a deadly contest of detection and deception in which defenders try to deprive attackers of the precious element of surprise. Instead of bullets, the combatants, both airborne and on the ground, fire waves of electromagnetic pulses in an escalating spiral of defensive measures, countermeasures, and counter-countermeasures. Rooted in the Second World War and the introduction of radar—for radio detection and ranging—electronic warfare has come to depend heavily on digital computers to control the lightning-quick moves of the game.

The United States military invests about $5 billion a year in electronic countermeasure equipment, and the Soviet Union is believed to spend a similar amount. The United States Air Force's B-1B strategic bomber, for example, is outfitted with the most complex electronic warfare system ever built—weighing two and a half tons and costing $20 million—to protect it against the Soviet Union's elaborate air-defense network. Even fighter planes, which have traditionally relied on speed and maneuverability, no longer can survive without electronic defenses.

The digital masterminds of this contest have many duties, from challenging an intruder for an electronic password to maintaining a library of past encounters that may provide some critical piece of information that can be used against the enemy. As the examples on the following pages demonstrate, Winston Churchill's wizard war now might better be called a battle of computers.

Radar's Mission: Detect and Track

The primary mission of military radar is to provide early warning of approaching aircraft and missiles. It does this by generating a fan-shaped beam of extremely high frequency radio waves that rotates continuously to illuminate a full hemisphere (below) every few seconds. As the radar's antenna turns, it transmits hundreds of high-power pulses of radio energy per second.

When the packets of electronic energy strike an aircraft, a portion of them are reflected back in the direction of the radar antenna. The radar listens for these echoes and determines the distance to the target by measuring the time that has elapsed since the pulse was sent out. Radio waves travel at the speed of light—roughly 1,000 feet per microsecond (a millionth of a second). Thus an echo received after an elapsed time of 500 microseconds reveals a target about fifty miles away (allowing

for two-way travel time). The direction in which the antenna beam is pointed when the echo is received indicates the target's direction, or bearing.

Once a target has been detected, it can be assigned to another kind of radar that tracks the intruder and aims defensive weapons at it. To pinpoint the position of the target, this tracking radar (shown as the inner bubble below) needs great accuracy, which is provided by a narrow, pencil-like beam of emissions. Tracking radar operates at higher frequencies, allowing it to generate a narrow beam with a moderately small antenna.

Computers contribute to this air surveillance in a multitude of ways. Among other roles, they are used in early-warning radars to distinguish between spurious reflections—off the ground or nearby mountains, for example—and echoes from real targets. They focus attention on any aircraft that fails to transmit an electronic code when interrogated by radar—indicating that the intruder probably is hostile. And on short-range tracking radars, computers provide the guidance to keep the beam aimed continuously at a designated target.

Alerting the enemy. Outbound radar emissions *(broad bands, above)* are always stronger than the echoes that rebound from the target to the radar transmitter *(narrow bands)*. Thus an aircraft with a radar-warning receiver can discover that it is being illuminated by radar before the radar detects the aircraft's presence. Warning systems on modern warplanes use a computer to analyze the radar signal in terms of its frequency, pulse repetition rate, and antenna beam motions. The resulting "fingerprint" is compared with those of known enemy radars, stored in a file memory, to determine whether this one is an early-warning radar or a short-range tracking unit, posing an immediate threat, and what countermeasures will be most effective.

Long-range detection. The inbound airplane above has flown into the radar's effective range—the distance at which the electronic echoes are strong enough to be detected. The limits of this range depend on the power of the radar's transmitter and how the target reflects the energy—a function, in turn, of the size of the aircraft and the angle at which it is viewed by the radar. Once received, the signals are analyzed by the radar's computers, which identify the target, determine where it is, and assist human operators to decide what should be done about it.

Electronic Tricks to Blind a Radar

Just as land and sea battles have often hinged on hiding movements from enemy eyes, so air attacks may involve efforts at concealment. One early method—still used against unsophisticated radars—is to have aircraft release millions of tiny aluminum-coated fiber glass slivers called chaff. Because chaff reflects radio energy, it can obscure targets from simple radars much as ordinary clouds can hide an aircraft from human eyes. A radar can tell the approximate location of a threat from a chaff cloud's position, but it cannot identify or count the targets, or determine their position accurately.

The Allies introduced chaff as a countermeasure to radar during World War II. Within months, German scientists responded with a counter-countermeasure that at least partially

A Screen of Chaff

A lead fighter plane called a pathfinder ejects a cloud of chaff that masks the planes behind it from the probing beams of long-range radar *(yellow bands)*. Chaff-dispensing systems use a computer to time the release, taking into account the velocity of winds aloft, the mission, and the type of radar involved. The result on the radar screen *(above)* is a blur in the general direction of the attackers.

neutralized the ploy. Their solution made use of a phenomenon first described by Christian Doppler, a nineteenth-century Austrian physicist. According to the so-called Doppler effect, the frequency of an echo from an approaching or receding object changes slightly from the originally transmitted frequency. This shift is proportional to the object's speed.

German radars were soon outfitted with filters that had the ability to separate the echoes of fast-moving aircraft from those of slow-moving, windborne clouds of chaff. Today's radars employ a similar technique, using banks of filters, controlled by computers, that are sensitive to the slightest differences between echoes.

A more advanced and effective method of neutralizing radar is noise jamming. Because radar depends upon receiving a rather weak echo from a target, it can be overwhelmed if an aircraft transmits a strong, continuous signal on the radar's operating frequency. The result can be compared with trying to carry on a conversation when a jet plane is taking off directly overhead.

Noise jamming, too, has its limits. Jamming transmitters require considerable electric power, which is costly to generate aboard an aircraft and is usually in short supply. Moreover, a jammer works on only a limited range of frequencies. To counter a variety of radars requires that an airplane carry many different jammers, which inevitably reduces either the plane's payload or its effective range.

Jamming with Noise

The fighter at lower left has come within range of enemy radar, but it neutralizes the radar by sending out a powerful, continuous signal (shown as a solid beam) that washes out the return echo and creates a distinctive V-shaped wedge on the radar screen (above). A computer on the jamming plane determines when to best invest the limited supply of electric power that creates the signal, what frequencies to use, and if there is more than one radar, which to jam first.

Exploiting Radar's Weak Flanks

One electronic countermeasure exploits the fact that all but the most recent antennae used for early-warning radar have a shortcoming. In addition to generating a fan-shaped primary beam, the radars also produce unwanted weaker emissions, known as sidelobes, on either side of the main beam. Normally the sidelobes cause no serious problems because, at longer ranges, they produce only very weak echoes that are readily distinguished from the stronger echoes of the main beam. However, an aircraft with suitable jamming equipment can exploit the sidelobes to disguise the plane's whereabouts.

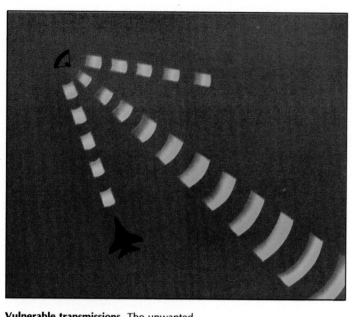

Vulnerable transmissions. The unwanted sidelobe beams generated on either side of the main beam (above) by a radar antenna makes it susceptible to mimic pulses returned by target aircraft. Radar is designed to select the strongest echo, on the assumption it comes from the main beam, and to ignore weaker sidelobe echoes. But if an aircraft transmits strong mimic pulses when it is illuminated by a sidelobe, a radar's tracking computer will assume that they come from the main beam. This deceives the radar as to the target's true bearing.

The deception process begins when the aircraft's radar warning system determines the angular displacement of the questing radar's sidelobes—typically about five degrees on either side of the main beam. An onboard computer measures the scanning rate of the ground antenna to predict when one of the sidelobes will be pointed directly at the plane. At that exact instant, the computer instructs the jamming system to generate a radarlike pulse that is identical to the real echoes—but much stronger.

The ground radar's computer, designed to track the strongest echo, is misled and begins to track the false pulse. This can lead to a substantial error in tracking. If the target plane is 200 miles away, the five-degree deception in bearing corresponds to a distance of seventeen miles.

As a result, the early-warning radar's computer may hand off the target to the wrong tracking unit, one whose range does not cover the area the aircraft is actually in. Or fighter aircraft scrambled to intercept the enemy plane will be misdirected, providing precious time that could allow the intruder to elude air defenses.

Chasing a Ghost

In the illustration at left a fighter plane sends spurious signals, shown in green, along the sidelobe of a distant radar. The radar mistakenly thinks it is picking up a return from its main beam (indicated by the ghosted plane at upper right). The screen above shows where the plane actually is in relation to the signal the radar receives (green blip). The longer the range at which this countermeasure is used, the greater the error it can generate.

A High-Stakes Game of Mimicry

Seconds can be crucial in the life-and-death duel between computerized air defenses and attacking aircraft. Many ground-to-air missiles are guided to their targets by radar, and if the radar loses track of the incoming plane even briefly, the missile may miss its target or run out of fuel, or the attacker may slip inside the defense to strike its objective.

A plane can sometimes confound a tracking radar by taking advantage of the radar's narrow focus. When the radar is

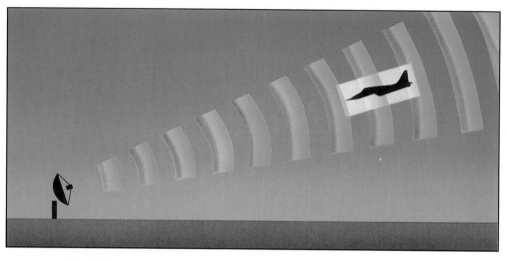

A discriminating radar. The tracking radar above has locked onto an approaching attack plane. To maintain a fix on the target, the radar acknowledges only echoes that come from the fighter's immediate vicinity *(box)*.

assigned a specific target, its computer registers the target's range and bearing; it then looks for echoes in that same area, while ignoring those from other ranges or bearings. Meanwhile, the plane's computerized radar-warning unit monitors the tracking radar's characteristics, and its jammer emits a strong pulse that imitates that of the radar. Because it is designed to track the strongest signal, the radar follows the phony echoes instead of the real ones.

Once it has the radar's attention, the plane creates a false range reading by delaying its transmissions, making the plane seem farther away than its true position. After the radar's tracking function has been lured a few miles off target, the aircraft turns off its jammer and the plane's image disappears from its expected position. At this point, the radar's human operator must intervene to try to find the lost target, and by the time the plane is relocated, it may be too late.

A Successful Deception

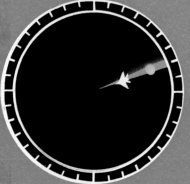

The attacking plane above, by responding to the radar's beam with strong mimic signals of its own *(shown in green)*, has managed to draw off the radar, which is still looking for it in the area indicated by the box. The ghosted plane represents the position of the target as mistakenly reported by the radar. The effect of this range deception is shown by the images on the screen at left: a blip where the plane is thought to be, and a silhouetted outline, closer in, where the plane actually is.

Radar's Rebuttal to Sham Signals

In the give-and-take of electronic warfare, ground-based radar enjoys certain intrinsic advantages. Attacking warplanes, no matter how sophisticated, are limited in the size and weight of their equipment and the amount of power they can generate for countermeasures. Ground-based radar installations do not suffer from such constraints, and normally they can generate very high levels of radio-frequency energy.

One result of this imbalance is that when an aircraft comes within a certain range of the radar, the strength of the radar's echo will be stronger than any jamming signal the plane

Radar's Brute Power

The radar installation at right takes advantage of its greater capacity to generate radio energy *(yellow bands)* to burn through the jamming signal of the approaching aircraft *(green bands)*. At this point the radar screen *(above)* accurately shows the planes' range and bearing from the transmitter, despite some background noise generated by the jamming.

can deliver. At that point, called the burn-through range, a radar operator is able to detect the oncoming target. The distance at which burn-through occurs depends on a number of factors, such as the target plane's radar cross section, the power radiated by its jammer, the peak power of the ground radar's transmitter, and the sophistication of its computerized signal processing.

Another counter-countermeasure available to ground radar is frequency hopping. Changing the frequency at which the radar transmits can frustrate the range- and bearing-deception techniques described on the preceding pages, in which enemy aircraft create pulses that mimic those of the radar. The true echoes return to the radar system while the false ones come in on the wrong frequency.

The more rapidly the radar can change frequencies, the more effective this defense will be, and the ability to select new frequencies at random makes it even more difficult to imitate. But if recent history is a guide, even this technique will be countered by some new offensive measure in the ever-escalating game of electronic hide-and-seek.

Frequency Hopping

The radar at left uses its ability to change frequencies in order to maintain a fix on an enemy fighter. The plane is trying to deceive the radar by emitting pulses that mimic those of the radar, but because the plane does not know which frequencies the radar will broadcast next, the signals it generates are ineffective. As a result, the true echoes are picked up by the radar, and an accurate sighting appears as a blip on the screen above.

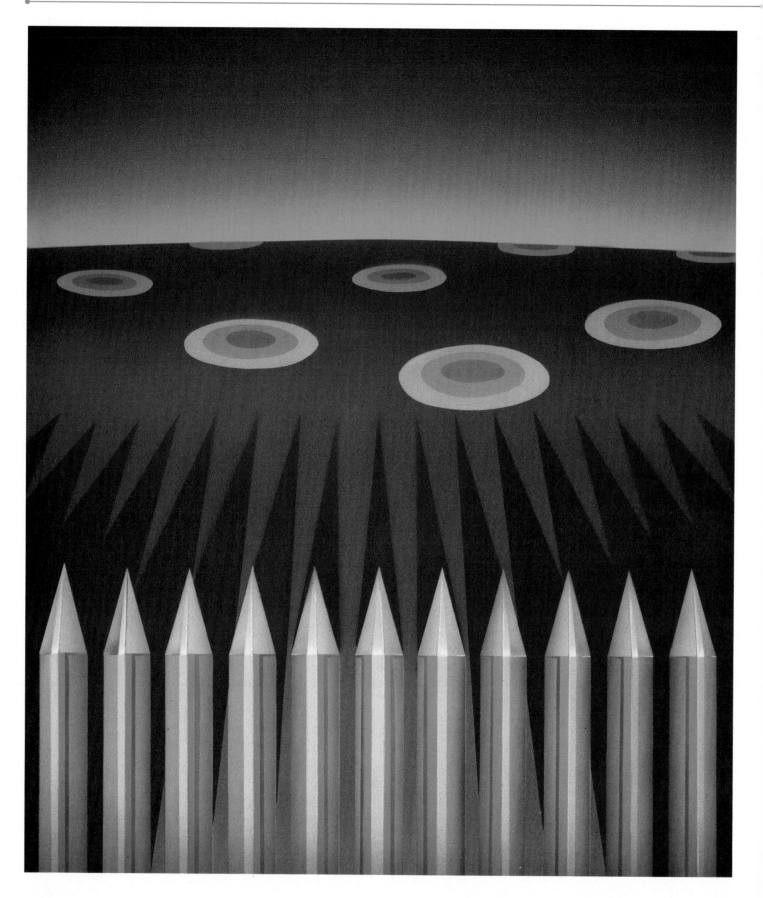

A New Age
in Weaponry

About four hours before dawn on April 15, 1986, eighteen American F-111 bombers approached the coast of Libya, flying at low altitude to evade radar. Their mission was to destroy suspected training sites for international terrorists. Because of the raid's political sensitivity and the fact that the targets were located in and around densely populated parts of the capital city of Tripoli, the planes were armed with the Paveway II, a laser-guided bomb.

Unlike its World War II predecessors—"dumb" weapons that, once released, were uncontrollable and more often than not fell wide of the mark—this modern bomb was designed for precise, surgical strikes. One of a growing arsenal of "smart" weapons, it was fitted with sensors and computer-controlled, movable fins capable of guiding the bomb to within a few feet of its objective. Each of the F-111s carried four of the 2,000-pound Paveways tucked under its wings.

The Pentagon's targets in Tripoli were the al-Azziziya barracks, home of Libyan leader Colonel Muammar Gaddafi; the Sidi Bilal port facility, where the Libyan navy trained divers; and the city's main airport, which was used by military aircraft. As each F-111 streaked toward shore, the two-man crew's weapons officer activated a Pave Tack electronic navigation and targeting pod mounted beneath the plane's fuselage; the pod contained an infrared television camera and a laser. Once the general shape of a target appeared on an F-111's radar, the infrared camera showed warm spots as black and cool spots as white. This served to further distinguish the target—whose heat signature had been previously recorded by intelligence-gathering infrared sensors—from the objects that surrounded it.

As the bombers hurtled toward the city about 400 feet above street level and traveling at more than 500 miles per hour, each weapons officer used a joystick to shine a laser beam on the target assigned to his bomber. Electronics in the Pave Tack pod then took over to keep the beam locked on the target. A computer calculated the optimal release point for the bombs, taking into account the velocity of the aircraft, the distance to the target, and other factors. At exactly the right instant, the computer loosed the Paveways from the wings. As the bombs fell away, a "seeker," or sensor, in the nose of each weapon detected the laser energy reflected from its target. A computer behind the seeker then commanded the bomb's movable fins to guide the weapon toward the center of the reflection.

The pilots had been ordered not to drop their Paveways unless they could positively identify the targets and execute a perfect low-level bombing run. The rules of engagement further stipulated that each target be identified by means of a double acquisition—that is, the target had to appear both on the planes' radar and on their Pave Tack infrared systems. However, because of Tripoli's "low radar significance"—Pentagonese for the lack of highly radar-reflective objects in the city—some crews were unable to meet these requirements and refrained from dropping their bombs. Furthermore, the crews encountered heavy surface-to-air missile fire. Although radar jamming by U.S. Navy jets offshore made the missiles erratic, they nevertheless forced the F-111 pilots to deviate from the nearly straight-and-level bombing run necessary for Pave Tack to achieve maximum accuracy. In some cases, the bombers' evasive maneuvers dislodged the laser beam from the target, leading the bombs astray.

At the al-Azziziya barracks, for example, Colonel Gaddafi escaped unharmed. Over the port of Sidi Bilal, meanwhile, clouds and smoke made the use of lasers uncertain, and the target suffered only superficial damage. Several military transports were destroyed at the airport, yet the toll at Tripoli also included private houses and a damaged French embassy. Bombs were found to have landed as far as two miles away from their targets. Although a World War II-style bombing raid would undoubtedly have killed more civilians, the raid on Tripoli was not exactly surgical. The mission suggested anew that the increasing computerization of modern military arsenals may contain equal parts wisdom and wishful thinking.

A TWOFOLD PATH

The quest for munitions that can distance men from the hell of hand-to-hand combat is as old as warfare itself. Mobility, armor, and firepower have all evolved over the centuries to give soldiers the hope, at least, that they can kill before they are killed. Yet the greater the range between adversaries, the more elusive a telling blow becomes.

Until World War II, all long-range weapons—rocks, arrows, bullets, cannonballs, bombs—were ballistic. That is, once a weapon had been released, its course could not be adjusted: It was determined by the acceleration of gravity as influenced by the height, angle, and speed of the missile at its release, and also by the wind and a long list of other factors. In other words, accuracy was all in the aiming.

By the end of the war, great strides had been made in the area of aiming. To help predict the trajectories of shells fired from the sixteen-inch guns of battleships, for example, the U.S. Navy fitted the vessels with electromechanical analog computers. In addition to the more obvious aiming factors, these computing devices considered the rotation of the earth and even the temperature of the propellant (the warmer the gunpowder, the faster it burns and the farther it lobs the shell) to calculate how the flight of each round would be affected. The Allied and Axis air forces used a variety of optical bombsights to help bombardiers hit targets. Premier among them was the U.S. Air Force's coveted Norden bombsight, which had a mechanical calculator that computed bomb trajectories, allowing for the plane's speed, altitude, and drift. As good as the Norden sight was, the results it achieved illustrate the limits of aiming. According to the U.S. Strategic Bombing Survey conducted after the war, only about 20 percent of bombs targeted for areas less than 1,000 feet across managed to strike home.

In the later years of the war, Germany and the United States began building guided weapons, some of which found their way into combat. In 1943, for example, Luftwaffe planes used radio-guided glide bombs, such as the Hs-293 and the Fritz X, to attack Allied shipping in the Mediterranean. An airman kept track of both the falling bomb and the target, sending course-correction commands to the bomb by radio remote control. On the Allied side, the Weary Willie, an unmanned, battle-scarred B-17 bomber directed by radio and carrying ten tons of explosives, flew missions against German submarine pens on an island in the North Sea in 1944 and 1945. These remote-controlled weapons had a negligible effect on the outcome of the conflict, however.

Nor did such weapons find much practical application in the immediate postwar years. The superpowers instead focused their research on improving the aiming capabilities of long-range bombers and on developing intercontinental ballistic missiles (ICBMs), which had limited course-correcting abilities, applicable only during the missiles' brief period of powered flight. By the early 1970s, however, emphasis had begun to shift away from better aiming toward continuously guiding a projectile toward the target after it left the launcher. The renaissance of this concept was made possible by rapid advances in two technologies—computers and target-sensing devices. Switching from vacuum tubes to transistors in electronic gear and the miniaturization that resulted from the invention of the integrated circuit—a thin slice of silicon overlaid with a microscopic network of transistors—made it possible to fit powerful computers inside weapons.

To give the computers something to steer for, a variety of sensors has evolved, capable of pinpointing targets even under adverse conditions. Infrared (IR) sensors, for example, are sensitive to light of slightly longer wavelength than the deepest red. Any object having a temperature above absolute zero emits small amounts of infrared radiation; as the temperature rises, the infrared radiation increases. Because this radiation penetrates fog and smoke, infrared sensors can detect tanks, trucks, diesel generators, and communications vans—all prodigal sources of heat—that might pass unnoticed by a sensor sensitive only to visible light. Compact radar transmitters also play a role in finding and marking targets. Used singly or in teams, these sensors and others make eluding detection much more difficult than ever before.

Computers and sensors, working hand in glove, made possible the first generation of weapons that could steer themselves (after a little assistance from foot soldier, seaman, or aviator) toward the target. Because of the electronic brain inside, these devices came to be called smart weapons, to distinguish them from their dumb, computerless ancestors. The earliest such armaments saw action in the waning days of the war in Vietnam and later in the sharp clashes that punctuate military life in Middle Eastern countries. Within a decade, the next generation—so-called brilliant weapons, capable of finding their way unassisted to the target—was making its debut. A soldier, in theory, can fire such weapons and then forget about them, free to dive for cover or to ready the next round. Some military theoreticians suggest that the ultimate step in the trend toward automated combat will be to populate the forward lines of battle with robot weapons manipulated by human controllers in the rear.

The principles behind smart and brilliant weapons are easy to describe in broad strokes, but applying the concepts is invariably a daunting proposition. If the hot exhaust system of a tank stands out to an infrared sensor, so does the intense flame of a flare that might be fired as a decoy. Designing and programming a computerized weapon to ignore the diversion and attack the tank is no small feat. As for hardware, it is little less than astonishing that integrated circuits, gyroscopes, and laser detectors can be built to withstand the rigors of the battlefield, including the shock of being fired from a cannon.

Small wonder, then, that smart and brilliant weapons are expensive, take a long time to develop, and sometimes fall short of expectations. The promise and problems of these weapons have sparked a lively debate over their utility, with one side calling for an arsenal altered in the direction of more numerous, simpler, and less expensive devices, and the other side steadfastly maintaining that such is the road to defeat in the next contest of arms. History suggests a likely outcome to the argument: In the scissors-cut-paper, rock-breaks-scissors world of military measures and countermeasures, technology has rarely moved backward. On the other hand, high tech has certainly stumbled badly at times.

COMPUTERS FOR FIRE CONTROL
Of all the threats faced by tanks and armored personnel carriers, enemy helicopters launching guided missiles from nearly two miles away are regarded by army planners as perhaps the most serious. To counter that menace—as well as to engage fixed-wing aircraft and even targets on the ground if necessary—the Division Air Defense gun was conceived. DIVAD, as it came to be known, is an example of good intentions gone awry.

DIVAD was intended to accomplish its mission through better aiming. To minimize costs and to speed the gun from drawing board to arsenal, planners decreed that the weapon be assembled primarily from off-the-shelf components. The result, unveiled in 1981, was two 40-mm cannon, made by the Swedish company Bofors, mounted in a turret on top of an M48 tank chassis. A search radar on the turret would detect targets, and a computer would then analyze the radar returns to establish the level of threat that each target represented: A radar reflection typical of a helicopter, for example, might trigger the highest alarm. The twin guns were to be aimed by a fully automatic, digital fire-control system based on a radar adapted from the air force's F-16 fighter. As the guns were trained on the top-priority target, the computer would select from two magazines the one loaded with ammunition appropriate to the target. Aerial threats would be engaged with proximity-fused shells designed to explode in a hail of shrapnel upon approaching within several yards of the target; armored personnel carriers and other targets on the ground would be fired upon with munitions designed to burst after impact. The DIVAD commander and gunner, after instructing the system to open fire, would monitor the action on a video display.

Although the DIVAD gun looked good on paper, it proved unable to hit its targets with any reliability during field tests. The main problem stemmed from the designers' insistence on using a radar intended for fighter-to-fighter air combat to aim a ground weapon at helicopters. The F-16 radar mated to DIVAD is a Doppler radar—a detection apparatus in which the increased frequency of a radar return indicates an approaching target, while a decreased frequency denotes a receding target. Doppler radars work fine against a jet fighter, which appears to the radar to be moving in only one direction. Against a helicopter, however, such a device is useless. As the rotor whirls, one blade appears to a Doppler radar to be approaching while another appears to be receding. What is more, the rate of approach or recession varies according to what part of the blade the radar returns are coming from—faster for returns from the blade tips and more slowly from sections near the hub. The effect is to make a hovering or low-speed helicopter all but invisible to the radar.

DIVAD fared somewhat better against fixed-wing aircraft, such as jet fighters, as long as the target stayed well above the horizon and refrained from maneuvering. But for a low-flying plane, radar reflections from the terrain became jumbled with those from the target. Inside the weapon's electronics, confusion reigned over where to aim the gun—at the ground or at the aircraft—with the result that DIVAD often fired between them.

After eight years of development at a cost of about $1.8 billion, DIVAD was abandoned in 1985, to be replaced by inexpensive, infrared-homing missiles such as Stinger and Rapier. Although it was an expensive lesson in how computerized weaponry can falter between design and production, DIVAD constituted a rare detour in the trend toward computerized weapons.

A more successful high-tech solution to the problem of aiming a gun is the laser range-finder system installed in the Army's M-1 Abrams main battle tank. The M-1 carries a computer that helps the gunner aim and fire the tank's cannon by analyzing information from several different sources. The process begins when the gunner fixes the cross hairs of his combination optical and infrared gun sight on the target—an opposing tank, for example—and activates the laser range finder to measure the distance to the target. To elevate the gun, the fire-control computer combines the range with the tilt of the turret (obtained from a sensor on the roof) and with a variety of data entered by keyboard before the battle: the type and temperature of the ammunition to be fired, barrel wear, the barometric pressure outside, and other variables. The computer, meanwhile, also notes the direction to the enemy tank from the position of the gun sight and causes the turret to swing the gun toward the target. Having aimed the tank's cannon, the fire-control computer displays a ready-to-fire message for the gunner, who then presses the trigger. Tanks of earlier generations were required to come to a complete halt in order to shoot accurately; the Abrams turret, however, is kept level by gyroscopes, permitting effective fire while the tank is under way.

A tank stands out against surrounding foliage in this computer-generated image from a laser radar (LADAR). To produce the picture, reflections of laser pulses, beamed at the scene, are used to measure the reflectance of objects within the system's field of view. LADAR provides sufficient detail for potential targets to be distinguished from objects such as trees or large rocks, a key step in making weapon systems that can identify and attack a quarry without human assistance.

A LANTIRN TO LIGHT THE WAY

A fighter pilot has no option but to shoot on the run. Against targets on the ground, he must do so accurately, on the first pass. A second try, made without the element of surprise and in the face of smart air-defense weapons, virtually guarantees that the aircraft will be hit, and probably destroyed. Under the best of conditions, spotting a target and loosing a missile or bomb at it while flying at high speed close to the ground threatens to exceed the capacity of even the best pilots. Yet pilots who fly planes like the General Dynamics F-16 are counted on to accomplish such feats on mission after mission, at night, or when low-hanging clouds obscure the view of the ground. Without computer assistance, the assignment would be impossible.

For these reasons, air force planners resolved to equip the F-16 and other aircraft intended for use against buildings, bridges, and other targets on the ground with a device that would allow them to operate day or night, regardless of the weather. The result is LANTIRN, for Low-Altitude Navigation and Targeting Infrared for Night.

First produced in 1981, the electronics for LANTIRN are packaged in two streamlined pods that fasten to the belly of the F-16. Each pod contains its own control computer working in conjunction with a master computer on board the aircraft and various microprocessors throughout the system, which overall uses about 200,000 lines of computer-program code—a relatively small number, given the complexity of the computers' tasks.

One pod contains equipment that permits an aircraft to fly 500 miles an hour through the dark just 200 feet above the ground. A forward-looking infrared sensor projects an image of the topography ahead of the aircraft onto a transparent cockpit screen, called a head-up display (HUD), that is positioned at eye level. The pilot uses the infrared image to survey the ground ahead for possible targets and places where targets might be hidden. So that the pilot can give his undivided attention to the task of searching out the enemy, terrain-following radar in the pod confidently guides the jet over hills and into valleys without risk of a crash.

The second pod, called the targeting pod, is designed to pinpoint stationary targets. It carries an infrared sensor of its own, as well as a combination laser range finder/target designator that is similar in function to the Pave Tack apparatus used in the raid on Tripoli. Over the target area, the infrared sensor in the pod scans the ground below and alerts the pilot to unusual temperature patterns that might signal the presence of a target, such as an enemy compound. Given the go-ahead by the pilot, the pod then automatically shines a laser beam on the target and releases a bomb or fires a missile that homes on the beam's reflection.

LANTIRN's shortcoming—one that is shared by most smart weaponry—is that the targeting system is incapable of interpreting the subtle temperature differences it records. It cannot distinguish between a parked truck, say, and some other source of heat, or between an enemy truck and a friendly one. The problem lies in the infrared system's inability to see a target in sufficient detail to positively identify it. The key to sharper vision is an advanced computer technique known as parallel processing, in which an array of 250,000 individual microprocessors would work simultaneously to analyze, thirty times a second, an equal number of pixels of an infrared CCD detector (page 38). The resulting acuity would make a Russian T-72 tank appear sufficiently different from a West German Leopard tank, for example, that computers could take over the targeting job from the pilot.

Flying a combat plane equipped with a pod capable of automatic target recognition (ATR), as this capability is called, a pilot would "pop up" from low-altitude flight in order to give the targeting pod a good view of the ground. During this brief exposure to enemy defenses, the LANTIRN pod would quickly pinpoint multiple enemy targets and would fire at all of them with the pilot's single squeeze of the "pickle"—the weapons-release button mounted atop the aircraft's control stick. That goal came a step closer in 1987 with the development of a prototype ATR computer that could perform 400 billion operations per second—more than 200 times as many as the general-purpose Cray 2 supercomputer—while nestled in a box only ten inches square and six inches high.

UPPING THE ODDS OF A HIT

Though intended for use against fixed objectives, systems like LANTIRN may one day be capable of dealing with moving targets such as columns of armor—either to prevent them from reaching the battlefield by striking them deep within their own territory, or to attack them as they attempt to overrun defending forces. Regardless of how accurately air power is deployed against them, however, many of the tanks will survive. Until the 1970s, the tanks' main opposition on the ground would have been enemy tanks. Since then, however, new weapons have made the ordinary foot soldier an additional threat to be reckoned with.

During the 1973 Middle East war, more than 1,500 Arab and Israeli tanks were destroyed or disabled in a few days by gunfire and "intelligent" missiles. Chaim Herzog, a major-general in the Israeli Army, described battlefields littered with tank debris and crisscrossed with spider webs of fine wires. The filaments of metal were the remains of Soviet-built antitank missiles fired by infantrymen and steered to their targets by means of manually operated "joysticks" used to send electrical guidance signals to the missile over the wires, which uncoiled from the rear of each rocket as it leaped toward its target.

An American version of the weapon, the TOW (Tube-launched, Optically tracked, Wire-guided) missile, had first seen action in Vietnam a year earlier. The TOW substitutes a computerized optical sight for the joystick. Fitted with an infrared sensor, the sight makes the TOW more accurate than its joystick-operated counterpart. Instead of having to be a skilled rocket pilot, a gunner has only to keep the sight's cross hairs trained on the target as the weapon flies through the air. The sight, which encompasses both target and missile in its field of view, tracks the missile by means of an infrared beacon in its tail. Steering commands pass from the sight through the uncoiling wire to a computer aboard the missile, which adjusts steering vanes as necessary in order to achieve a direct hit.

The TOW and its cousins have shortcomings, chiefly that they are relatively slow and require the missileer to keep the target in view until it is struck. The TOW takes fifteen seconds to cover a distance of 3,000 yards. This could be enough time for an alert tank crew to move the vehicle to cover or to fire in the direction from which the missile was launched. Even a near miss might force the soldier who fired the TOW to take cover and lose control of the projectile, or shrapnel might sever the all-important guide wires.

Nonetheless, hundreds of thousands of TOW missiles and their Russian cousins have been sold around the world since their impressively destructive performances of the early 1970s. Indeed, one school of military thought holds that the continuing refinement of antitank weaponry may soon render the tank obsolete. Other theoreticians contend that countermeasures will be found. In the early 1980s, for example, the Soviets began draping their main battle tanks with explosive steel plates called reactive armor. Based on an Israeli design, the plates add as much as two tons to the vehicle weight and are designed to explode

Machines for Dangerous Jobs

Computerized drones of various types are able to eliminate the human risk encountered in carrying out some of the most critical wartime missions. Combat intelligence, which was once gathered by courageous individuals operating behind enemy lines, can be more successfully ferreted out by remote-control airborne cameras. Patrolling the perimeter of a military installation, an activity that exposes human sentinels to surprise attack, can be handled by computerized guard vehicles that are extraordinarily difficult to sneak up on. With a computerized minisubmarine, for another example, every major vessel in the navy can have the advantage of its own mine-sweeping service.

Each of the drones shown on these pages is equipped with an onboard computer, which is preprogrammed to perform specific maneuvers and tasks. In addition to reporting discoveries made by a variety of onboard sensors, the devices respond to commands from human operators; these commands are communicated by radio link or, in the case of the Pinguin (bottom), by fiber-optic cable.

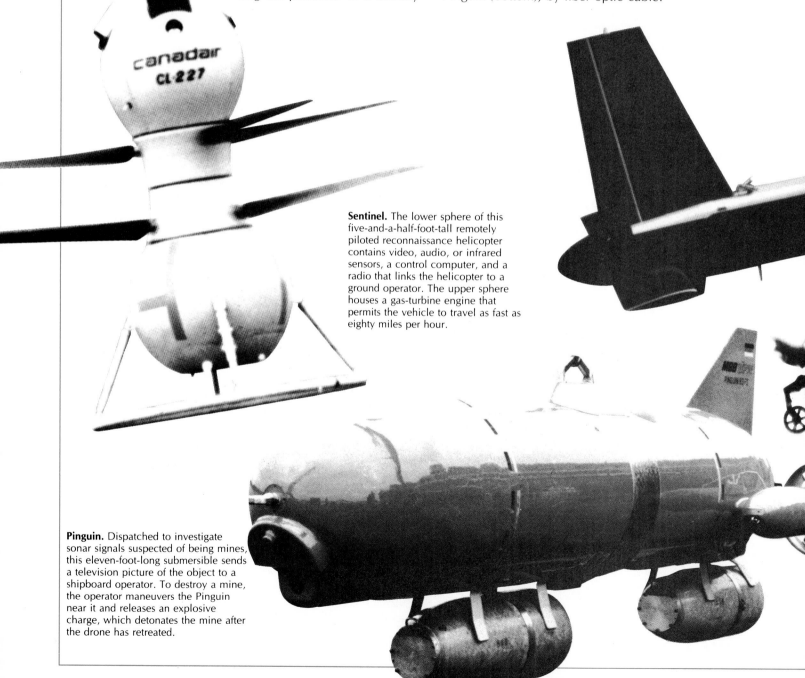

Sentinel. The lower sphere of this five-and-a-half-foot-tall remotely piloted reconnaissance helicopter contains video, audio, or infrared sensors, a control computer, and a radio that links the helicopter to a ground operator. The upper sphere houses a gas-turbine engine that permits the vehicle to travel as fast as eighty miles per hour.

Pinguin. Dispatched to investigate sonar signals suspected of being mines, this eleven-foot-long submersible sends a television picture of the object to a shipboard operator. To destroy a mine, the operator maneuvers the Pinguin near it and releases an explosive charge, which detonates the mine after the drone has retreated.

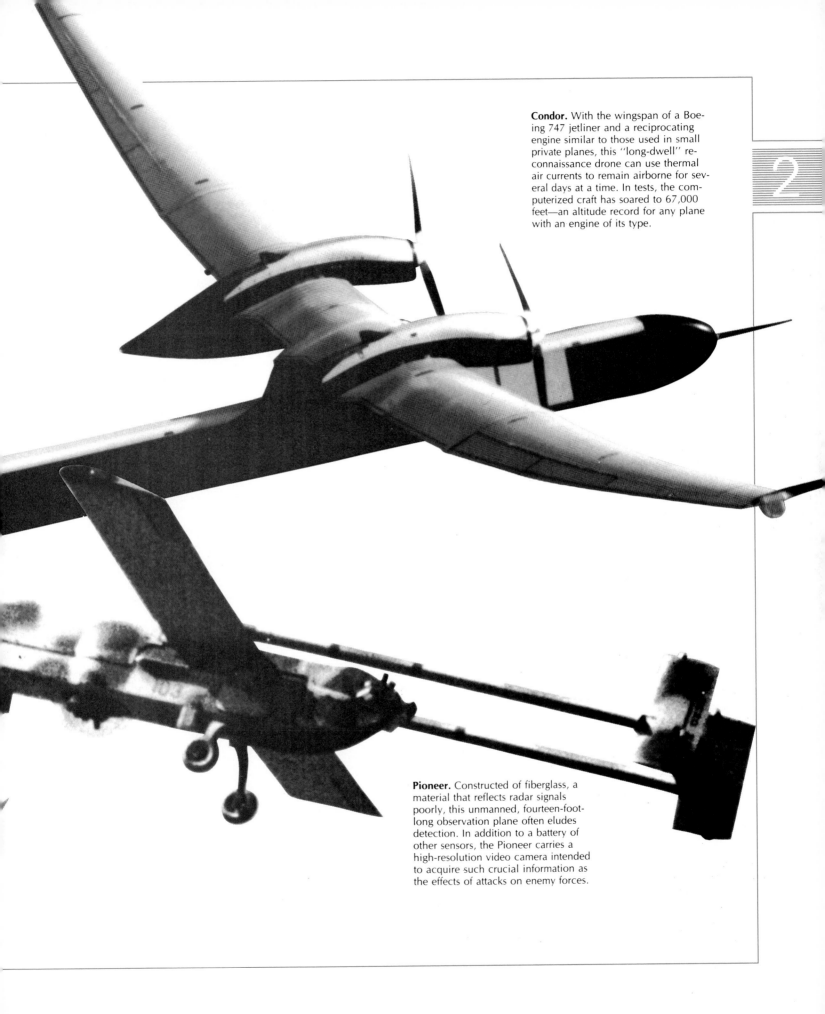

2

Condor. With the wingspan of a Boeing 747 jetliner and a reciprocating engine similar to those used in small private planes, this "long-dwell" reconnaissance drone can use thermal air currents to remain airborne for several days at a time. In tests, the computerized craft has soared to 67,000 feet—an altitude record for any plane with an engine of its type.

Pioneer. Constructed of fiberglass, a material that reflects radar signals poorly, this unmanned, fourteen-foot-long observation plane often eludes detection. In addition to a battery of other sensors, the Pioneer carries a high-resolution video camera intended to acquire such crucial information as the effects of attacks on enemy forces.

when they are hit by the shaped charge of an antitank missile, rendering the missile ineffective but keeping the tank intact.

IN HARM'S WAY

Using radar to guide weapons relieves the gunner of steering duties. Such devices respond to the intensity or some other characteristic of radar emissions. The first missile to apply this principle used an enemy radar's own emissions to home on the source of the signals. Named the Shrike, it was first fielded early in the Vietnam War to neutralize surface-to-air missile (SAM) and antiaircraft artillery (AAA) sites around Hanoi and other targets inside North Vietnam. A receiver in the missile picked up enemy radar signals, using them as a beacon to home on the antenna at supersonic speed and destroy it.

North Vietnamese antiaircraft crews soon learned to befuddle the Shrike simply by shutting off their radars, depriving the missile of a signal to home on. The countermeasure's countermeasure, embodied in a second-generation

missile called the Standard ARM (antiradiation missile) fielded in 1968, was to put a computer aboard. The computer constantly plotted a course for the missile to the target's last calculated location, to be used in the event that the radar went off the air.

Antiradar missiles reached a lofty stage in the High-speed Anti-Radiation Missile, or HARM. Deployed with first-line squadrons in the mid-1980s, the HARM is considerably more versatile than either the Shrike or the Standard ARM. Against a target having a known location, for example, the HARM is provided with the range and direction to the target from a computer aboard the launch aircraft. As the missile nears the target, the radar seeker turns on and the HARM homes on the radar transmitter. Traveling at more than three times the speed of sound, the missile is likely to find its mark before the radar crew can turn off the transmitter. The HARM is usually used in this manner by fighters providing cover for strike aircraft. The launch is timed so that the missile arrives at the target at the same time as the attacking aircraft, when the enemy radar is most likely to be turned on.

The missile is also used in a so-called surveillance mode. In this role the HARM keeps a sensitive electronic ear cocked for enemy radar signals. When a signal is detected, the missile is launched. Supplied with information about the wavelengths and other characteristics of enemy radars, the HARM can be prepro-

Images from a television camera in the nose of a Walleye self-guided bomb, displayed on a small monitor in the cockpit of the aircraft that dropped the weapon, show the missile nearing its target, a bridge near Thanh Hoa, North Vietnam. Before dropping the Walleye, the pilot pointed the bomb at the target, bringing the desired point of impact within a square formed by two pairs of dark lines *(far left)*. Then the pilot released the weapon. A computer aboard the Walleye steered the missile to the target, keeping the point of aim within the square. A post-strike reconnaissance photograph *(opposite)* reveals the bomb's devastating accuracy.

grammed to track down as many as three different types of radar, weighing priorities according to the threat presented by each. Should the radar posing the gravest danger shut down, the HARM automatically shifts its attention to the second-priority target, and so on. If the first radar returns to the air, it will again attract the missile's attention.

The HARM and its predecessors are examples of "fire-and-forget" weapons, which require no human attention once they have been released. Over the years, research for such weapons has focused on improving the "sight" of their guidance sensors—that is, on giving the weapons the ability to distinguish their targets. Toward the end of the Vietnam War, for example, the U.S. Navy deployed a bomb that was guided by television. Dubbed the Walleye, it contained a 2,000-pound charge and was fitted with a television camera in its nose, a computer, and steering vanes. Before releasing the Walleye, the pilot flew at the target, which the television camera displayed on a cockpit monitor. The computer recorded the television image as a pattern of light and dark areas, and

stored this digital map in memory. As the pilot pickled the Walleye and veered away, the onboard computer steered the bomb toward the target by issuing guidance commands that kept the view through the camera lens aligned with the light map in its memory.

DEATH FROM A SEA-SKIMMER

Far better known than the Walleye is a French-built fire-and-forget missile, the Exocet. An antiship missile (the name means "flying fish"), the Exocet was already ten years old when first used in combat during the Falkland Islands war in 1982. One of the missiles, fired from an Argentine warplane, gored the British destroyer *Sheffield* and exploded in a fireball belowdecks, touching off a conflagration that claimed the lives of twenty British sailors and eventually forced the scuttling of the ship.

The Exocet is known as a sea-skimmer, meaning that, for most of its flight, it flies just over the wave tops, lurking below the radar horizon of the ship it is attacking. Even when the Exocet enters the sweep of a ship's defensive radars, it presents a small radar cross section—the profile that an object engenders in a radar system as it is tracked. Finally, the Exocet's low avenue of attack makes it difficult for a defender to separate the missile itself from the radar images created by the wave crests just below it. The result is a negligible interlude between detection and impact. As the *Sheffield's* captain James Salt recalled, "We had time only to say 'Take cover!' "—about three or four seconds—before the missile hit the ship.

The Exocet's unnoticed approach was made possible by a nose-mounted radar altimeter that enables the missile to fly just six feet above the surface of the ocean. The device calculates the missile's altitude by bouncing radar waves off the water and timing their return. This data is then fed to the Exocet's small onboard computer, which tilts the missile's tail fins to keep it flying at the minimum height.

Before firing the missile, the pilot of the launching aircraft must provide the Exocet's computer with such critical information as the direction in which the target is moving, the range, and the altitude at which the missile will be released. After launch, the missile stays on course by obeying an inertial guidance system. At a point about six miles from the target, a radar guidance system takes over, bouncing radar waves off the ship and zeroing in on the center of the reflections. Because the Exocet's radar can be jammed or disoriented, this final portion of the missile's flight is potentially its most vulnerable. But at that point, a ship has little more than thirty seconds to defend itself.

A ship's best hope for protecting itself against sea-skimming missiles may be to fire missiles of its own. Indeed, since September 1967, when the Israeli navy lost the destroyer *Elath* to three Egyptian navy Styx missiles of Soviet manufacture, arms makers the world over have spent billions of dollars to develop self-defense missiles that can engage sea-skimmers before they strike home. The most effective such weapons, called snap-launch point-defense missiles, are the Israeli Barak and the British Seawolf. The Seawolf, used by the *Sheffield's* sister ships in 1982 to bring down five Argentine aircraft on low-level attack runs, is so finely calibrated that it can hit a 4.5-inch artillery shell in flight.

WANTED: MIDAIR COLLISIONS

An air-to-air missile using radar guidance similar to that which steers the Exocet through its final moments of flight has proved somewhat more difficult to perfect, perhaps because planes are better able than ships to outmaneuver an attacker. By the late 1980s, the most sophisticated air-to-air missiles were the Israeli Python, the British Skyflash, the French Magic and Mica, and the American AMRAAM (Advanced Medium-Range Air-to-Air Missile). These missiles use their own radar to home in on targets beyond visual range, allowing a pilot to fire the weapon and immediately break away to engage another enemy threat.

Hughes Aircraft Company of Los Angeles, California, won the contract to develop the twelve-foot-long, 335-pound AMRAAM. To assess whether the missile would make a difference in the outcome of aerial engagements between fighters, Hughes designers put computers to work. A large computer model named TAC Brawler *(pages 16-17)*, containing 130,000 lines of code, simulated air-to-air combat involving numerous aircraft, some armed with AMRAAMs, some not. The creators of TAC Brawler used a form of artificial intelligence to fashion plausible combat scenarios featuring such real-life aspects of dogfighting as surprise, confusion, and cooperation among pilots. Although TAC Brawler increased the designers' confidence that the AMRAAM was worth building, computer simulations of any military hardware are far from definitive. Even the best models are simplifications of reality and may hold out more promise for a new weapon than is justified. In the case of the AMRAAM's, the confidence seemed justified, though many doubts were raised during nearly a decade of development.

Perfecting the complex electronics of the AMRAAM turned out to be a particularly thorny task for the missile's manufacturer, and fitting the components into the missile posed significant difficulties. The AMRAAM measures just seven inches in diameter, yet inside its sleek casing nestle eight circuit boards, six electric batteries, four miniature electric motors, a radar system, a fifty-pound warhead, and a solid rocket engine capable of driving the missile through the air at twice the speed of sound.

AMRAAM had to pass some hard tests before getting the final go-ahead. For example, two missiles, launched at the same time, were required to knock a pair of target drones from a sky filled with chaff (page 52) ejected by an aircraft in midair to confuse a pursuer's radar. In one instance, the missiles' failure to differentiate between the two drones caused both AMRAAMs to chase down and destroy the same target. But in April 1987, a brace of AMRAAMs fitted with redesigned seeker and guidance systems successfully intercepted their targets, clearing the way for the missile's eventual production.

THE DEADLIEST ACCURACY OF ALL

Although most of the weapons that underwent computerization in the 1970s and 1980s had dumb ancestors, the intercontinental ballistic missile was an exception. No one expected that an ICBM lacking a guidance system could land anywhere near its target after a flight of thousands of miles. Even so, guidance systems permitted a missile to drift from its ideal course at such a rate that, when first fielded by the United States and the Soviet Union in the 1950s, ICBMs were expected to miss their targets—the centers of cities, industrial complexes, and military bases—by several miles. To ensure that no target would pass unscathed, the missiles carried nuclear warheads vastly more powerful than the atomic bomb dropped on Hiroshima in August 1945. Gradually, accuracy has been honed to a fine edge. One result is much smaller warheads, though they still handily exceed the yield of the Hiroshima bomb. The improved accuracy is largely the result of miniaturized electronic components, which allow more computer functions to be performed aboard the missile.

ICBMs of every vintage rely on inertial guidance units (IGUs) to get them close to the target. IGUs employ gauges called accelerometers to measure the effects of forces that influence a missile in flight, primarily thrust from the engines. But aerodynamic influences, such as winds aloft, also play a role. The accelerometers feed this data to an onboard computer, which uses the information to calculate the missile's velocity and position. Comparing the resulting figures to data stored in memory about where the missile should be according to the flight plan, the computer then issues the commands necessary to make any steering adjustments.

An IGU is designed to ignore the effects of gravity, also an acceleration, in order not to mask the consequences of acceleration from other sources. Yet gravity is the most important influence on a missile's flight. The solution is to estimate gravitational effects in advance and store them in the computer's memory before the missile ever leaves its silo. Electronic miniaturization permits remarkably detailed approximations to be stored in the onboard computers, yet the information remains a projection. No ICBM can be flight-tested along the route it would travel to the Soviet Union, a course over the North Pole that would take a missile through regions where the earth's gravitational field tends to vary, with effects on the flight path that cannot be confidently predicted.

Within three minutes after launch, the IGU must have done its job well. By that time, the missile will have exhausted its fuel and can no longer be steered with the rocket exhaust. Moreover, it will have left the atmosphere, making steering vanes useless. For most of the remainder of its thirty-minute flight, an ICBM behaves like the dumbest of weapons. It is little more than a huge artillery shell, subject only to the laws of ballistics. A guidance error of one degree early in the flight can cause a warhead to miss its target by more than 100 miles.

The most sophisticated ICBM guidance instrument is the Advanced Inertial Reference Sphere (AIRS), built for the U.S. Air Force's MX missile by the Northrop Corporation of Hawthorne, California. Located just below the missile's nose cone, the AIRS is a marvel of ingenuity in a small space. No larger than a basketball, the AIRS has a shell of strong beryllium alloy. Packed carefully inside

A Beeline to a Target

Cruise missiles—flying bombs powered by small jet engines—are in many ways the ideal long-range weapon. They can be launched a safe distance from a target and are inexpensive and small enough—generally no more than eighteen feet long—to be carried on submarines, ships, and aircraft. Terrain-following radar permits them to frustrate defenses by flying less than 100 feet above the ground. Furthermore, they are astonishingly accurate: These missiles could split the uprights of a goalpost in a football stadium after flying 700 miles.

An onboard computer manages the three guidance systems necessary to assure such deadliness. The first is an inertial navigation system (INS). By keeping track of the smallest changes in the missile's speed and direction, an INS is accurate enough for a nuclear-tipped missile to destroy most targets. But against a blast-resistant target such as a missile silo—or for a cruise missile armed with a chemical explosive such as TNT—greater precision is needed.

To provide it, a system called Terrain Contour Matching (TERCOM) is used several times during a mission to correct INS errors. As explained overleaf, TERCOM compares the height of the ground passing beneath the missile to a computerized topographical map of the route. In the final minutes of flight, the missile shifts to a third system called DSMAC—Digital Scene-Matching Area Correlator. DSMAC matches views of the ground taken by a camera mounted in the missile with pictures of the approach to a target stored in memory: the result should be a bull's-eye every time.

A typical mission. Launched offshore, a cruise missile skims across the open sea, guided toward the coast of enemy territory by an inertial navigation system. Upon crossing the shoreline, TERCOM checks that the missile is on course, a procedure repeated at predetermined locations along the route. Terrain-following radar prevents the missile from crashing into the rugged terrain on the way to the target.

A Topographical Data Base

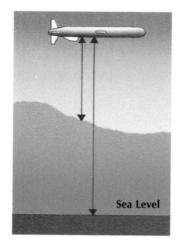

Measuring altitude. A cruise missile needs two kinds of altimeters to find the height above sea level of the terrain below. A radar altimeter measures the missile's height above ground by timing the passage of radio signals that it bounces against the landscape. A barometric altimeter measures air pressure to determine the missile's height above sea level. The missile's computer subtracts the radar-derived value from the barometric one to determine the height of the terrain.

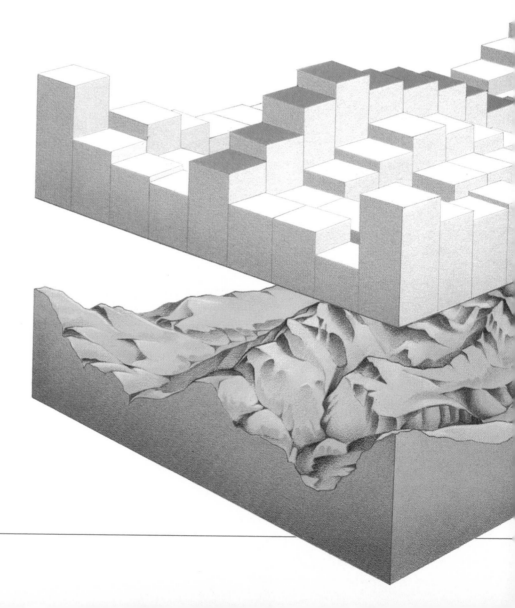

The data base of onboard topographical maps that a cruise missile's TERCOM system uses to stay on course is based on surveys by reconnaissance satellites that may employ radar altimetry to measure terrain height, much as the missiles themselves do *(left)*. The computer that controls TERCOM compares the average height of terrain with the topographical data. To be suitable for TERCOM, land must be hilly or mountainous. Otherwise, the average height of one square will not vary enough from others nearby for the system to detect the difference. Thus, the oceans and flat expanses of desert are unsuitable for TERCOM navigation. Equally inap-

propriate are areas where geographic ups and downs seem to repeat themselves for many miles. Rolling terrain, for example, where one hill is much like another, might confuse the TERCOM system.

During a mission, TERCOM is activated by the computer on cue from the INS, which estimates when the missile arrives within the coverage of a TERCOM map. As explained below, TERCOM detects any navigation errors that the inertial system has allowed and corrects the missile's course. Within a few miles of the objective, the photographic guidance system DSMAC takes over.

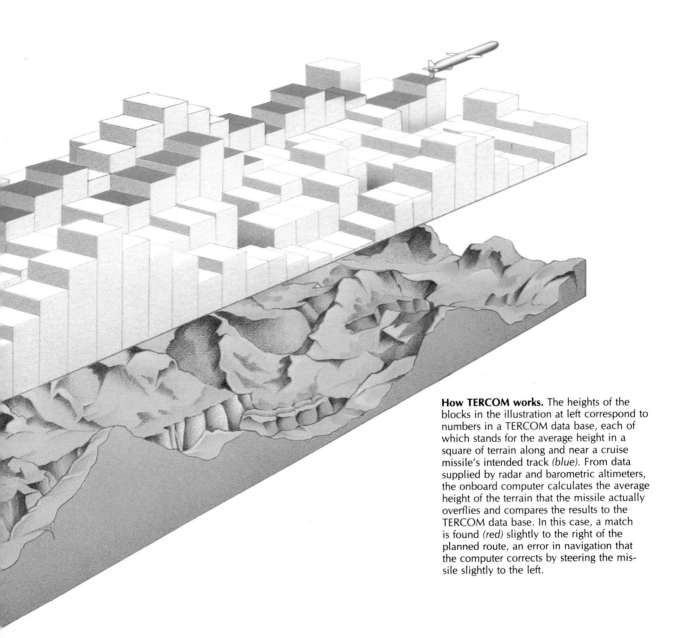

How TERCOM works. The heights of the blocks in the illustration at left correspond to numbers in a TERCOM data base, each of which stands for the average height in a square of terrain along and near a cruise missile's intended track *(blue)*. From data supplied by radar and barometric altimeters, the onboard computer calculates the average height of the terrain that the missile actually overflies and compares the results to the TERCOM data base. In this case, a match is found *(red)* slightly to the right of the planned route, an error in navigation that the computer corrects by steering the missile slightly to the left.

are 19,401 parts, ranging from gyroscopes to integrated circuits. Working together, they are expected to give an MX missile the ability to strike within several hundred feet of its intended point of impact.

DIGITAL CRUISE CONTROL

Inertial guidance provides an effective means for steering missiles that, like ICBMs, reach the target relatively soon after launch, but the drift from course allowed by even the best systems—a few tenths of a mile per hour—would make slower weapons such as ground-hugging non-nuclear cruise missiles unacceptably inaccurate. Flying at 500 miles per hour, a cruise missile might take as long as three hours to reach its target, enough time for it to miss the target by a mile or so. Military engineers and computer scientists therefore teamed up long ago to devise a scheme called terrain-contour matching, or TERCOM, which was intended to allow cruise missiles to penetrate an enemy's radar defenses by flying as low as sixty-five feet above the ground. By the 1950s, the U.S. Navy had readied for flight testing two prototype cruise missiles, dubbed Triton and Rigel, that featured TERCOM navigation and ramjet propulsion yielding speeds of Mach 2 to Mach 3.

Although the concept of terrain-contour matching is straightforward, its execution was difficult and took nearly three decades to perfect. TERCOM depends on teamwork between a computer and a radar altimeter, both housed inside the missile. The computer's memory holds digitized maps depicting the contours of the terrain at way points the missile will skim over during the course of its flight. As the missile reaches the approximate location of each way point, guided there by an inertial navigation system, the computer alerts the radar altimeter, which bounces radio waves off the ground. These readings, which show the missile's height above the ground, are combined with readings from an onboard barometric altimeter, which show the missile's height above sea level; together the two sets of measurements yield an actual map of the terrain below. The actual map is then compared to the map in memory, and the computer issues course-correction commands as necessary to bring the two into alignment (pages 75-77).

TERCOM has limitations. For example, sufficiently distinctive way points are rare in some areas of the world, making certain targets unsuitable for a TERCOM-guided attack or requiring a circuitous route to the target in order for the missile to pass over navigable terrain. Construction projects, landslides and even erosion can alter topography in ways that make it difficult for TERCOM to reconcile the terrain detected by its radar altimeter with the maps stored aboard the missile in programmable read-only memory (PROM) chips, a type of memory that, once filled with data, cannot be altered. If the system misses a single way point, the cruise missile becomes hopelessly lost. A relatively recent development in computer-memory technology may help to remedy the problem, however—and also provide greater flexibility in choosing targets. Called EEPROM (for electronically erasable programmable read-only memory), it is a variety of PROM that will allow the terrain maps resident in a cruise missile's computer to be updated just before the weapon is launched.

Perhaps TERCOM's most important limitation is that its accuracy, though a substantial improvement over an inertial navigation system operating unassisted, lacks the precision to give cruise missiles the capability of hitting a target as small as a communications van, for example, after a flight of several hundred miles. A navigation technique called DSMAC, for Digital Scene Matching Area Correlator, may hold the answer. The DSMAC approach is reminiscent of the TV-and-computer system that the navy used to guide its Walleye bombs in Vietnam. A photographic image showing the last few miles of the route to the target, acquired by satellite and stored in memory, is compared with a picture of the same landscape taken in flight. Sharing the computer used by the inertial system and TERCOM, DSMAC continuously corrects the weapon's course to reconcile differences between the two views.

A RAIN OF TERROR

Cruise missiles and ICBMs, jet fighters and bombers, tanks and antiaircraft guns are perhaps the most prominent examples of computerized weaponry, so much so that they overshadow a rich assortment of less publicized devices. One such weapon—an artillery round capable of homing in on several tanks simultaneously—illustrates how imaginatively computers can be employed to neutralize the hardware of war.

On the drawing boards for the 1990s, the device is called SADARM—Sense and Destroy Armor. Meant to stem an onslaught of enemy tanks or mechanized infantry, SADARM in one version is to be fired from a 155-mm artillery piece toward an assembly of tanks. A dumb high-explosive round from the same gun would have to hit a tank directly to destroy it—a lucky shot indeed.

A SADARM round, however, compensates for less-than-perfect aim. Inside the device are three or four submunitions that are released above the tanks. Each submunition is a canister that contains an explosive charge, a metal slug, an infrared sensor, and a radar that operates at a frequency of many billions of hertz, or cycles per second. The wavelength of this type of radar—as short as a few millimeters—yields a detailed image using an antenna less than six inches in diameter, small enough to fit snugly inside the artillery shell. A signal-processing computer completes the package.

Most probably, SADARM would be employed in large numbers against concentrations of enemy armor. Upon being liberated from the cannon shells over the target, the canisters drift to earth suspended from parachutes designed to descend in a spiral, permitting the radar and infrared sensor to survey the scene below. While the radar looks for a mass of metal, the infrared sensor searches for sources of heat. The computer compares notes from both devices. If a mass of metal coincides with a heat source, the computer triggers the explosive, which propels the metal slug through the thinly armored top section of a tank, an armored personnel carrier, or a self-propelled artillery piece.

Smart weapons like SADARM that use multiple sensors to find the target are thought to be extraordinarily difficult to defend against. Flares, used by aircraft to throw heat-seeking missiles off the trail, would not fool the radar aboard SADARM. Depending on SADARM's final design, jamming of the radar might cause the signal processor to give more weight to data from the infrared sensor, but a signal processor programmed with pattern-matching capabilities would make effective decoys a logistical nightmare. Not only would the lure have to be hot to flummox the infrared sensor, it would have to be roughly the size and shape of a tank. And even if half an armored division were converted to decoys, that measure alone would reduce SADARM's lethality by only 50 percent.

THE BATTLEFIELD OF THE FUTURE

A statement made in 1969 by General William Westmoreland, the commander of American troops during the Vietnam War, reflects a technological vision of intelligent weaponry that prevails to this day. "On the battlefield of the future," Westmoreland prophesied, "enemy forces will be located, tracked, and targeted almost instantaneously through the use of data links, computer-assisted intelligence evaluation, and automated fire control. I see battlefields that are under twenty-four-hour real- or near-real-time surveillance of all types. I see battlefields on which we can destroy anything we locate through instant communications and almost instant application of highly lethal firepower."

What Westmoreland had described, in effect, was a battlefield almost too deadly for a battle to occur. "On a heavily armed front, such as central Europe or the Middle East," wrote Paul Walker in 1981, "the side that strikes first, thereby giving away its position, will be the more vulnerable side." Field exercises tend to support this assertion, at least indirectly. As long ago as 1972, U.S. helicopter gunships assigned to NATO and firing TOW missiles each scored disabling "hits" on eighteen West German tanks, playing the role of aggressors, before being shot down. In 1980, helicopter crews participating in Operation Bright Star in the Egyptian desert confirmed the lesson. By flying close to the ground and taking cover behind hills, they reported, "we could shoot at 3,000 meters, then move back and shoot again. We could eat up enemy tanks as long as we had room" to fire at long range. The sun may have set on the Blitzkrieg.

Work under way on a Pentagon project called the Smart Weapons Program seemed to lend additional substance to General Westmoreland's prediction of a fully computerized battlefield. Under the auspices of the Defense Advanced Research Projects Agency (DARPA), the program aimed at nothing less than applying "advances in computer architecture and machine intelligence to the development of autonomous weapons capable of attacking high-value targets deep in enemy territory. These weapons will have the ability to analyze their environment and current battle situation, search likely target areas, detect and analyze targets, make attack decisions, select and dispense munitions, and report results."

The computerization of weapons may thus be a circular process, with each new advance in the electronic arsenal creating an environment that demands yet further computerization. Although the ultimate step of total autonomy for battlefield robots is unlikely, since the services would never consent to relinquish every shred of human control, the intermediate steps will no doubt continue to fuel the military's insistence on weapons with brains.

The Supercockpit

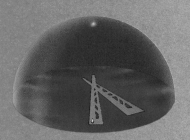

As the technology of powered flight approaches the end of its first century, combat aircraft threaten to exceed the capacity of humans to fly them. The Mach 2 F-16, for instance, can twist and turn about the sky in maneuvers that make the pilot feel nine times as heavy as his actual weight. These gyrations may either drain blood from a pilot's head, causing loss of consciousness called a blackout, or force excess blood into the skull, resulting in a state of sensory confusion called a redout.

Yet pilot alertness is more critical now than ever before. On a combat mission, a fighter pilot must combine a welter of instrument data and external information to create a three-dimensional mental image of his target, his aircraft, and the enemy's defenses. To help fighter pilots achieve this state of awareness under the stressful conditions of battle, the United States Air Force is developing a system of computerized assistance known as the supercockpit. Planned for the mid-1990s, it will present to the pilot all the crucial facts in a single, panoramic display.

Central to the supercockpit is a helmet equipped with a special optical system that superimposes, on the pilot's view through the canopy, pictures from an infrared television camera plus an overlay of computer-generated reports on such factors as altitude and airspeed, weapon and engine status, and distance to a target from a laser range finder. A computer-synthesized voice warns the pilot of approaching enemy jets or other hazards, and a cadre of intelligent subsystems offers him advice on how best to respond to threats. Thus—as exemplified by the imaginary mission on the following pages—the pilot gets all the information needed to fly and fight without having to divert attention from the situation developing outside the cockpit.

An Intimate Relationship

Embarking on a mission to destroy a bridge, a fighter pilot climbs into the cockpit and dons a custom-fitted helmet and a pair of special gloves, activating the supercockpit system. A data-transfer module, brought to the plane by the pilot, sends target, threat, and navigation information to the computer. The computer tracks the position of the aviator's head and hands by noting electrical signals induced in magnetic detectors in the gloves and helmet as they move within a magnetic field in the cockpit. Eye motion is measured by reflecting infrared light from his eyes into a micro-TV camera.

From the pilot's helmet orientation, infrared sensors aimed outside, and stored data, the computer generates an image to supplement the view through the cockpit. Consisting of topographical contour lines and other important landscape features, the image is projected onto the helmet's visor to help the pilot fly at night or when poor visibility causes the real world to dim. A hemispherical bubble in his lap shows a bird's-eye view of the scene, and a computer-generated image of a control panel displays functions necessary for each phase of a flight. The pilot issues commands by touching the control panel as if it were a real one, or by voice.

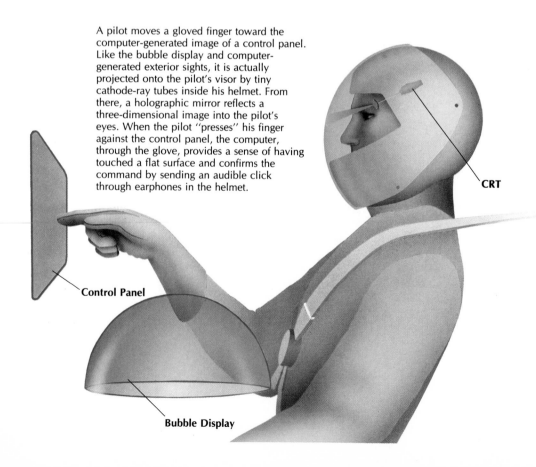

A pilot moves a gloved finger toward the computer-generated image of a control panel. Like the bubble display and computer-generated exterior sights, it is actually projected onto the pilot's visor by tiny cathode-ray tubes inside his helmet. From there, a holographic mirror reflects a three-dimensional image into the pilot's eyes. When the pilot "presses" his finger against the control panel, the computer, through the glove, provides a sense of having touched a flat surface and confirms the command by sending an audible click through earphones in the helmet.

CRT

Control Panel

Bubble Display

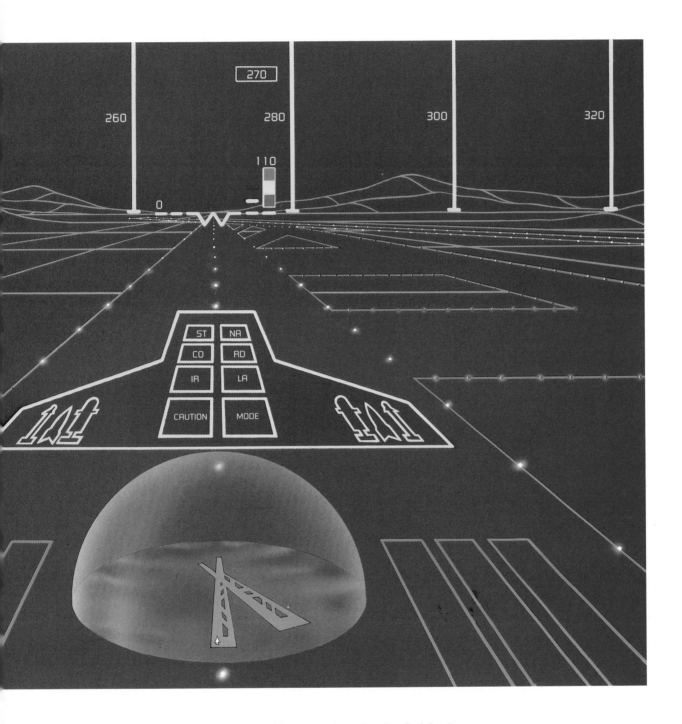

Yellow vertical lines in this computer-generated view indicate compass headings; the jet's heading, 270 degrees, appears in a box at the top of the picture. Altitude—110 feet above sea level—is displayed above colored boxes having a pointer that indicates whether the jet's present course is unobstructed (green), includes possible obstacles (yellow), or would certainly result in a collision if unchanged (red). The winged W shape nearby shows whether the plane's nose is pointed above or below the horizon and whether the plane is in level flight or turning. Airspeed appears to the left of the W. Grid lines accent critical topography and give the pilot a sense of speed as they flash by. Superimposed on the runway, a crude representation of the jet has symbols that allow the pilot to drop bombs or fire missiles. An eight-feature control panel lets the pilot display the jet's status, or vital signs (ST), and invoke navigation systems (NA) and communications (CO), as well as radar (RD), infrared (IR), and laser countermeasures (LA). The bottom two boxes are for warning the pilot of aircraft malfunctions and selecting a mode of operation—en route, air-to-air combat, and ground attack, for example. The bubble display shows the entire airfield, with the jet visible at the near end of the runway, poised for takeoff.

Driving to Work

After an uneventful takeoff at dusk, the pilot can relax; the onboard computer, linked to a network of sensors, closely monitors every aspect of the flight. A situation-assessment module, one of several subsystems in the computer, continually scans the skies for other aircraft. The systems-status module measures oil and hydraulic pressures, as well as engine and exhaust temperatures. If the module detects an unusual value, it alerts the pilot by lighting the CAUTION box on the control panel.

The computer also keeps tabs on the pilot. A module called the guardian is hooked to sensors that measure the pilot's blood pressure, breathing pattern, and pulse rate. These measurements would enable the guardian to recognize the physiological signs of blackout or redout and automatically take control of the aircraft until the pilot recovered. In addition, the computer would limit the aircraft's response to any instructions from the pilot that would stall the jet, put it into a spin, or otherwise damage it.

In the bubble display, the pilot sees the planned flight path, represented by a red ribbon. A small white wedge indicates the fighter's current position and direction. The evening's target, a bridge, is marked with a *T*. The pilot may observe any part of the flight path from a different angle or in greater detail. To inspect the terrain around the bridge, for example, the pilot could call up a magnified view by pointing to the target with an index finger and speaking the command "Zoom."

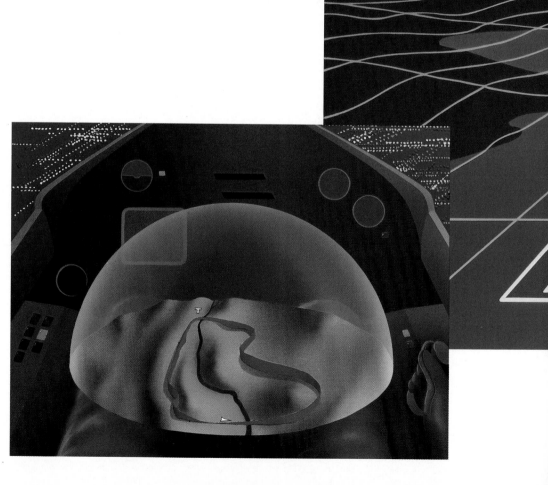

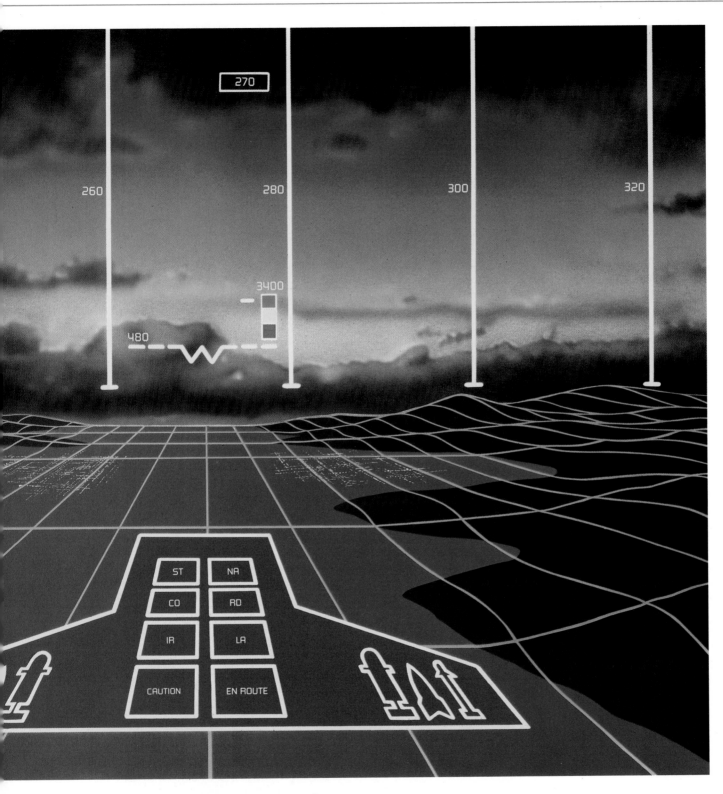

The jet heads into the sunset at a cruising altitude of 3,400 feet. As the sky darkens, computer-generated imagery will occupy more and more of the pilot's view, replacing virtually all of the real scene except lights on the ground, such as those that illuminate the two cities near the horizon.

Countering a Threat from the Air

As the jet flies down a canyon to elude enemy air-defense radar, the computer's speech synthesizer suddenly warns of unidentified jets approaching on the right: "Bogies, two o'clock." Simultaneously, symbols for two aircraft appear in the pilot's field of view. The situation-assessment module, linked to infrared search-and-track equipment, has detected the jets, and now, using the laser rangefinder, calculates range, heading, and speed.

Identifying the new players is the first order of business. The pilot says, "Identify IR," specifying that the computer use an infrared television camera instead of radar, which could reveal his presence. The computer displays an infrared image for the pilot, who immediately recognizes the distinctive features of enemy interceptors.

The situation-assessment module (which has also identified the jets by means of an intelligent pattern-recognition program) alerts the tactics module, which handles battle management. It evaluates and ranks threats against the plane, and tells the pilot when to fire weapons. Within a split second, the tactics module has evaluated the pilot's options and recommends a flight path that maximizes the chances of shooting down the enemy fighters.

The pilot may override the proposal, accept it, or ask for more information. In this instance, he follows the suggestion and instructs the computer to take over flying the plane so that he can concentrate on the kill.

The enemy jets are displayed in the upper right corner of the pilot's field of view, highlighted by a red circle. The mode display on the control panel has changed from ENROUTE to AIR TO AIR, reflecting the sudden shift to battle conditions. Snaking to the left is a series of three-sided boxes that show the pilot where to fly for the best shot at the enemy. Preparing to fire, the pilot touches the computer-generated control panel at the air-to-air missile mounted under the right wing and says, "Select." As the supercockpit computer lights the symbol to acknowledge the command, the pilot looks directly at the image of the closer of the two jets and says, "Aim." In response, the computer activates the missile's infrared guidance system, which locks onto the target.

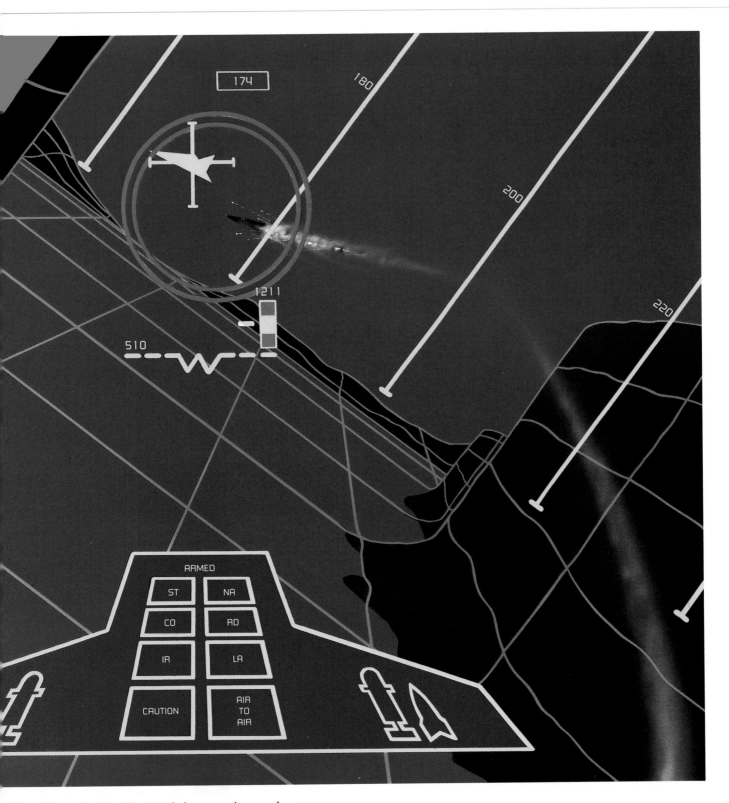

The missile, launched by a verbal command on cue from the tactics module, streaks toward the target. The pilot designates the other enemy jet as his second target just as the first one explodes in flames.

A Safe Route to the Target

Minutes after the successful air engagement, the pilot nears his target. The bridge is defended by a surface-to-air missile (SAM) site and a battery of antiaircraft artillery (AAA), guns that fire explosive projectiles. Colored balloons—white for SAMs and red for AAA—show the weapons' threat envelopes, or lethal ranges, which the situation-assessment module extracts from an onboard library of threat envelopes for each kind of antiaircraft weapon.

Balloons are positioned according to intelligence loaded into the computer before takeoff. To confirm the locations of enemy defenses—and update them if necessary—the computer analyzes signals emitted by antiaircraft radars as they search the skies for approaching planes. Each type of radar sends out pulses at a distinctive rate and wavelength. These characteristics allow the computer to identify both the radar and the weapon it serves.

As the attack begins, the tactics module plots the safest route between the two threats and displays it to the pilot. The pilot accepts the computer's suggested flight path and streaks down toward the bridge.

Three-sided boxes indicate the safest route to the bridge between a SAM site on the left and a AAA battery on the right. As the jet drops into its bombing run, the mode display on the control panel changes to ATTACK. When the pilot selects the television-guided air-to-ground missile mounted under the right wing, its symbol glows. Next, he designates the target with an "Aim" command. The tactics module waits for the optimal moment, then signals the pilot to launch the missile with a verbal "Fire" command.

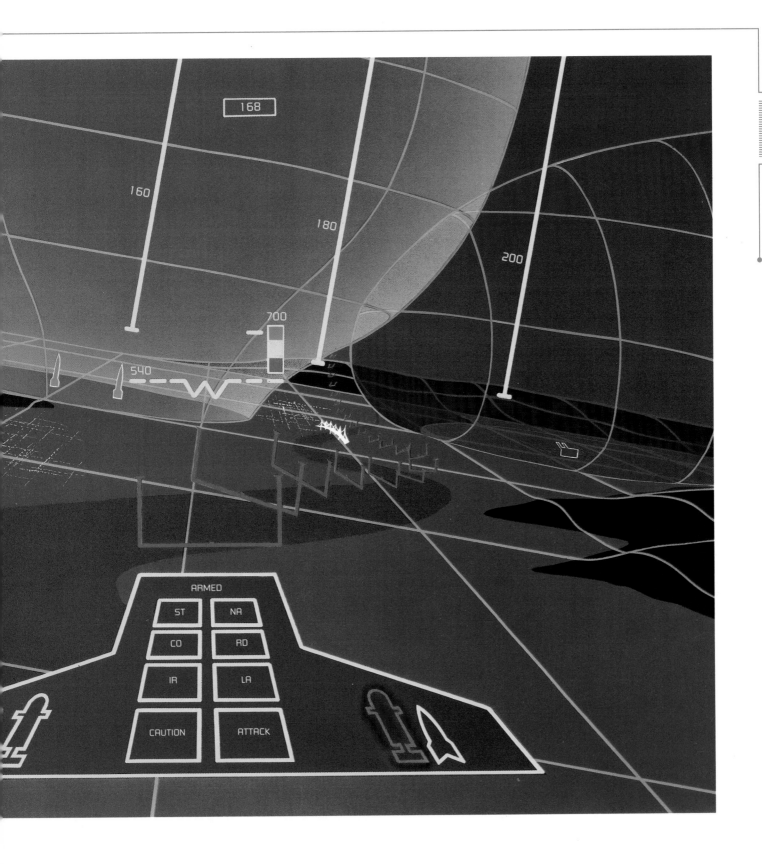

Limping Home, Tail Dragging

The pilot has achieved his objective; the bridge sustained a direct hit and collapsed. But an ambush by a AAA battery, which maintained radar silence until the fighter came within range, has damaged one of the fighter's two rudders, control surfaces on the tail that help the plane turn.

Onboard computers assess the damage, and within seconds, they have repositioned control surfaces on the wing, showing the new arrangement to the pilot in the display bubble *(below)*. Instead of struggling to compensate for the shrapnel-riddled surfaces, the pilot continues to fly the plane as usual. Each movement of the control stick is fed through a section of the computer called the reconfiguration module, which reinterprets the pilot's instructions to compensate for the damaged rudder.

Although the onboard computer eases flying, it cannot restore the fighter's overall maneuverability. As the lights of the airstrip come into view, the computer recommends a landing approach that is less demanding of the aircraft than one the pilot might fly if it were undamaged.

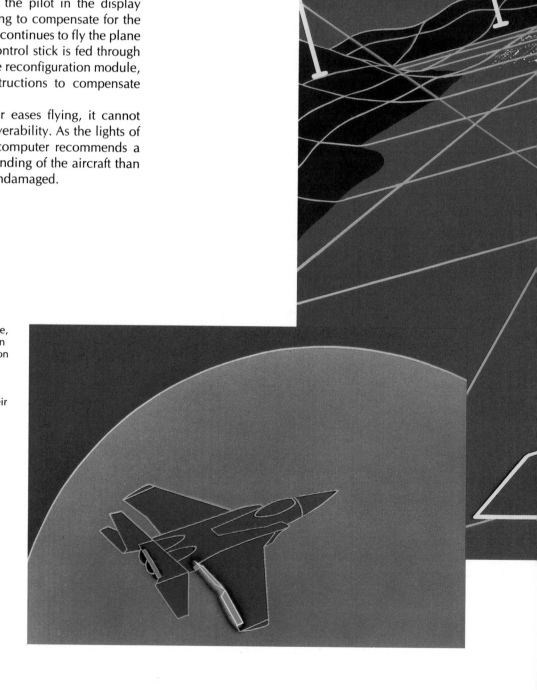

Seconds after being struck by antiaircraft fire, the pilot views the condition of his fighter in the bubble display. Sensors in the tail section have informed the supercockpit computer's reconfiguration module that the starboard rudder *(red)* is out of commission. Having been informed by other sensors through their associated computer modules that the jet now responds abnormally to the controls, the reconfiguration module compensates by deflecting the right flap and aileron, highlighted in green.

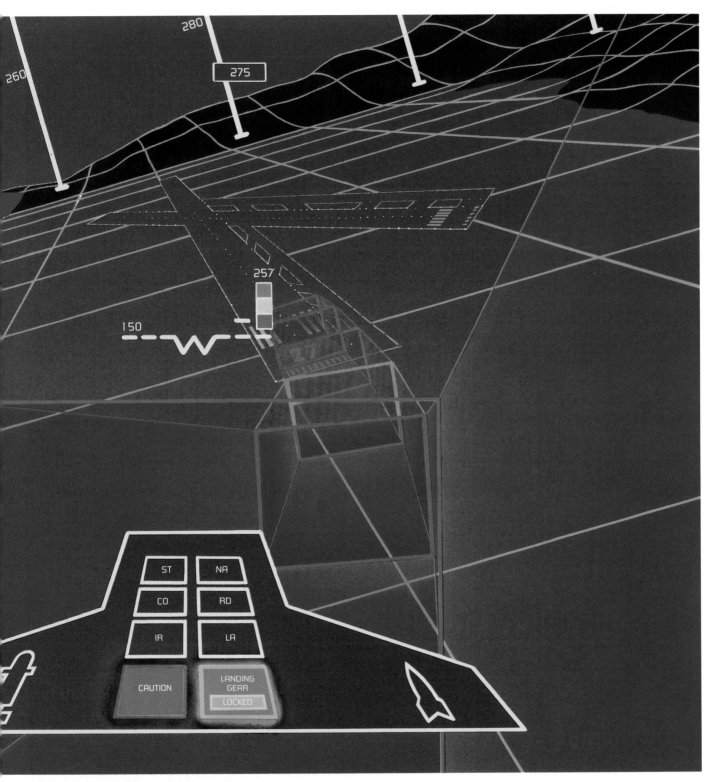

The glowing mode display on the control panel tells the pilot that the fighter's landing gear is extended and locked for landing; a caution light reminds him that the jet is limping home. A computer-generated flight path (closed boxes indicate that it is more restrictive than similar recommendations earlier in the mission) shows a safe descent to the runway.

The Computer and C3

3

The U.S. invasion of the Caribbean island of Grenada in October 1983 was much more difficult than it need have been. The 6,000 invaders, assigned to expel about 1,100 Cuban troops from the island, enjoyed clear superiority in arms as well as numbers. Yet breakdowns in communications plagued them from the outset. Army and marine forces approaching the island from opposite directions failed to establish radio contact, with the result that units from both services opened fire upon each other. Army commanders were unable to call in supporting fire from navy aircraft and ships, while other communications lapses put ground troops in jeopardy of attack from their intended air cover. In one instance, the ground forces fired at friendly planes.

Dismayed by the communications failures of the operation, the Defense Department took a variety of remedial actions. One of the first was to resolve a long-standing dispute between the air force and the navy over a computerized communications scheme known as the Joint Tactical Information Distribution System, or JTIDS. The interservice wrangling had centered on the relative emphasis to be given to voice communications (favored by the air force) and to data communications (favored by the navy), and the quarreling had kept JTIDS—which was conceived in 1975—from its planned deployment across all the U.S. services. Equipped with a system like JTIDS on Grenada, those services would have been able to converse in a common language and strike as a unified force.

THE NEED FOR C3

JTIDS's absence in 1983 and the improvements it promises to bring about in the future suggest how dramatically computer linkages can enhance the combat readiness of U.S. forces around the globe. Able to interconnect everything from strategic-bomber and ballistic-missile commands to infantry platoons, fighter squadrons, and solitary submarines prowling the ocean depths, computers form the central nervous system of the military physique.

Of all the ways in which computers help the military to flex its muscle—by pinpointing enemy forces, for example, or by refining the accuracy of the arms used to strike them—their emergence as tools for command has perhaps altered warfare the most. Computers have become crucial instruments in the hands of a commander for two main reasons. In the decades since World War II, the growing number and sophistication of intelligence-gathering sensors has vastly increased the amount of data a commander must take into account before he can make an informed combat decision. At the same time, the development of highly computerized "electronic arsenals"—coupled with brain-boggling advances in the speed and effectiveness of the vehicles used to deliver their lethal contents—has drastically reduced decision-making time.

Only computers can sift through the avalanche of pertinent information fast enough to allow the quick responses that modern warfare demands. In so doing, computers have become the key elements of an elusive military ideal known as command, control, and communications—C3, or C-cubed in Pentagon argot.

Over the years, the task of defining C3 has generated nearly as much controversy as have the various means proposed to achieve it. Air force colonel Kenneth Moll suggested the scope of the problem in 1978 when he observed that C3 is "subject to wildly different interpretations. The term can mean almost everything from military computers to the art of generalship." The field of rival descriptions has since narrowed, and now most strategists agree that C3 can be defined as the process by which commanders decide on combat actions, set them in motion, and direct their courses. So basic are computers to the process that C3 is sometimes referred to as C4—command, control, communications, and computers. In both the tactical (battle-fighting) and strategic (war-fighting) realms, computers are everywhere. They orchestrate weapons systems designed to perform jobs ranging from the protection of a clutch of ships at sea to the space-based destruction of nuclear missiles flying halfway around the world.

A PORCUPINE AFLOAT

It was a computerized ship-defense system called Aegis (named after the mythical shield of Zeus) that signaled the beginnings of the Pentagon's move toward automated command and control. The navy began developing Aegis in 1974 to protect its aircraft carriers and their escorting warships from antiship missiles and aircraft that might penetrate the pickets formed in the sky by the carriers' interceptor aircraft.

The U.S. Navy's need for Aegis was brought home in the late 1970s, when Admiral Sergei G. Gorshkov presided over the transformation of the Soviet Navy from a coastal defender to a worldwide strategic presence. The thorn in the club of the Soviet Navy's buildup was an anticarrier program of missile-equipped surface ships, submarines, and aircraft—among them, land-based Backfire bombers fitted with antiship cruise missiles capable of demolishing an enemy vessel more than 350 miles away. Analysts believe this combination of forces could muster aerial assaults so overwhelming that only a fast, automated system such as Aegis could defend against them.

By 1990, fifteen cruisers fitted with the Aegis system had been built for service as part of U.S. Navy carrier battle groups. Thanks to the system's ability to detect, identify, and track a multitude of enemy targets—be they incoming missiles, surface vessels, or hostile submarines—Aegis promises to make navy warships less vulnerable than ever before, despite the dramatic increase in the threat to these vessels.

Aegis works its protective magic through a combination of a phased-array radar *(pages 28-29)*, sixteen Unisys UYK-7 mainframe computers, twelve Unisys UYK-20 minicomputers, and a variety of defensive weapons. The Aegis radar, incessantly adjusting the phases of its 4,000 electronically radiating elements, can simultaneously track as many as 280 incoming missiles. Whether a foe could marshal his attacking forces with enough precision to deliver that many missiles at the same time remains a subject of debate among naval strategists.

The Aegis system's command-and-decision computers correlate readings not only from the cruiser's own radar and sonar, but also from similar sensors aboard other ships in the battle group. The command-and-decision computers then transmit this data to the weapons-control computers, which rank the targets according to the level of threat they represent and fire the appropriate weapons. All the while, the Aegis computers exchange information through an automated radio data link with the computers aboard other ships in the battle group. These ship-to-ship communications include such critical information as the identity of hostile and friendly traffic in the area, the targets being engaged, and the targets most likely to be engaged next.

To deal with airborne targets, the Aegis system has at its disposal as many as 122 fifteen-foot-long Standard (SM-2) antiaircraft missiles, weighing 1,556 pounds each; it can direct roughly a sixth of these interceptor missiles in the air at a time. Once the missiles are away, the weapons-control computers tell guidance microcomputers in the missiles which targets to head for. (Targets may be as distant as eighty-five miles or as close as four, and they are vulnerable at altitudes up to 80,000 feet.) The missiles then track signals bounced off their targets by a shipboard transmitter called an illuminator and close in for the kill by homing on those reflected signals. A third and final set of computers monitors data generated by the computers in the command-and-decision and weapons-control sets and calculates the probability of a missile intercepting its target.

For the close-range engagement of missiles that elude the interceptors, the Aegis computers also oversee the firing of two six-ton, six-barreled Phalanx Gatling (or machine) guns mounted on the cruiser's deck. Each Phalanx gun spits out 20-mm uranium-core bullets at the rate of fifty per second, throwing up a metal curtain between the cruiser and the incoming missile. (Uranium is used because it is two and a half times denser than steel, giving the bullets that much greater punch.) Computers in each Phalanx system track the trajectories of the bullets as well as that of their onrushing target, adjusting the gun's aim until it pulverizes the intruder in midair less than one mile from its intended victim.

An Aegis cruiser can prove hazardous to an enemy's submarines as well, although enemy subs are considerably more difficult to detect than hostile aircraft and antiship missiles. A typical Aegis cruiser bristles with an array of advanced antisubmarine weapons that make conventional depth charges seem antiquated by comparison—although the cruiser carries these, too. Mark 46 acoustic homing torpedoes are fired from one of the vessel's two torpedo rooms, whose hatches open just above the waterline, while antisubmarine rocket torpedoes (ASROCs) are fired from the deck. The ASROC, a deadly refinement of the Mark 46, flies through the air in the direction of the target's last reported vicinity, jettisons its rocket engine, and descends to the surface of the sea by parachute. Once the ASROC is in the water, it uses active sonar to pursue the submerged quarry.

The cruiser can also carry two Seahawk SH-60B LAMPS (Light Airborne Multi-Purpose System) attack helicopters fitted with their own Mark 46 torpedoes. From the cruiser, the choppers range far out to sea in search-and-destroy missions aimed at sinking enemy subs first detected by the cruiser's sonar.

The Aegis cruiser is also menacing to surface ships. It carries a pair of automatic-loading deck guns that can lob five-inch shells a distance of twelve miles once every three seconds. Even when an enemy vessel lurks beyond that range, however, it may be justified in fearing an Aegis cruiser. Awaiting launch from steel tubes on the cruiser's afterdeck are eight Harpoon antiship cruise missiles, American cousins of the French Exocet. Each Harpoon employs an

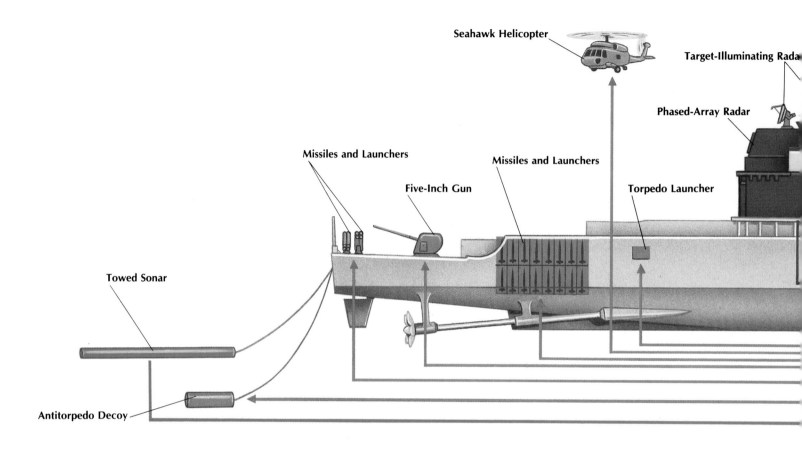

Seahawk Helicopter

Target-Illuminating Rada

Phased-Array Radar

Missiles and Launchers

Missiles and Launchers

Five-Inch Gun

Torpedo Launcher

Towed Sonar

Antitorpedo Decoy

onboard guidance computer to deliver a 500-pound conventional or nuclear warhead to targets that may lie over the horizon. A solid-fuel boost rocket blasts the Harpoon free of its firing canister, whereupon a turbofan jet engine takes over for long-range, wave-top cruising. During the missile's final attack run, its guidance computer issues a chain of commands that cause the missile to take evasive action, climbing from the surface of the water to dive down upon its target.

The automation of naval combat has prompted a parallel streamlining in the shipboard fighting force. In contrast to the 100 officers and nearly 1,700 sailors that the less-potent missile cruiser of the past needed aboard, the navy can man an Aegis cruiser with just 25 officers and 330 sailors.

The toughest part of getting Aegis operational lay in bringing the software up to par. Twenty-eight computer programs govern the system, and those with the job of telling the command-and-decision computers how to recognize and track airborne targets experienced chronic problems. The most nettlesome was the computers' tendency to confuse targets that had crossed paths. Devising the software algorithms to solve this and related problems added to the program's expense. By 1984, the Aegis component of each cruiser accounted for about one third of the vessel's $1.1 billion cost.

A RADAR-COMPUTER MÉNAGE
Similar software difficulties bedeviled an air force program known as AWACS, for Airborne Warning and Control System. Its object was to create

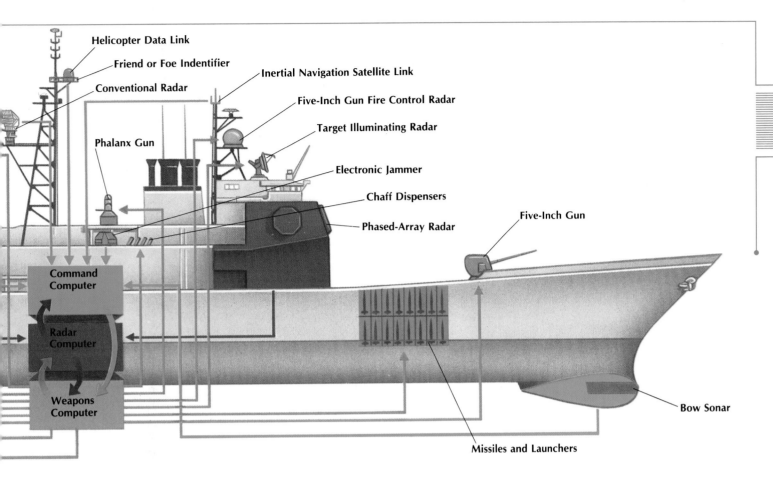

Helicopter Data Link

Friend or Foe Indentifier

Conventional Radar

Phalanx Gun

Inertial Navigation Satellite Link

Five-Inch Gun Fire Control Radar

Target Illuminating Radar

Electronic Jammer

Chaff Dispensers

Phased-Array Radar

Five-Inch Gun

Command Computer

Radar Computer

Weapons Computer

Bow Sonar

Missiles and Launchers

Intended to shield itself and neighboring ships from attack, the Aegis cruiser is among the most computerized of American fighting vessels. At the heart of this potent cruiser's defenses lie three interconnected computer systems. The radar computer is dedicated to the ship's phased-array radar *(red)*, which can track hundreds of airborne targets simultaneously *(pages 28-29)*. Information thus gathered is passed to the command computer, which also processes data from sensors such as sonar and conventional, rotating-antenna radars *(green)*. In action, the command computer uses this information to assess the threat posed by each attacker—ship, submarine, aircraft, or missile—and orders the weapons computer system to employ the cruiser's arsenal of guns (including two Phalanx Gatling guns that each fire 3,000 rounds per minute), missiles, and torpedoes *(blue)* against the intruders, beginning with the most threatening.

a fleet of flying command posts that would manage aerial warfare as Aegis is used to manage sea battles. AWACS planes, the first of which entered service in 1977, are modified Boeing 707s crammed with computers and display consoles. The aircraft are designated E-3A, E-3B, or E-3C depending on the number of crew stations they contain and the sophistication of their target-tracking radars. An E-3 typically carries fourteen AWACS specialists in addition to its flight-deck crew of four.

All E-3s can chart the progress of airborne attackers at distances of nearly 400 miles, and then vector, or guide, friendly fighters to intercept them. Their radar antennae are mounted in huge, rotating domes (called roto-domes) atop the fuselages. The radars must be able to pick out and pursue low-flying targets—including cruise missiles—among a welter of earth-reflected, radar-return signals known as ground clutter.

The AWACS program was barely eight years old when the air force realized that the progressively smaller cross sections, or radar profiles, of Soviet fighters and cruise missiles might soon permit such threats to escape the system's notice. Not only that, reported air force general Charles Cabell in 1985, but "We face more jamming today than we did in 1977. We expect to find more jamming in the 1990s than we do today. The Russians recognize the value of the E-3 and the need to defeat it."

Those prospects spurred the air force to begin enhancing the ability of AWACS radars to detect smaller targets and resist enemy jamming. The air force set out to accomplish this not by outfitting the aircraft with new radars, but by upgrading the AWACS signal-processing computer, which receives and processes the radar signals, and a second computer, called the radar-data correlator, that analyzes those signals and controls the radar. The data correlator's computational speed will be at least quadrupled, to roughly ten million operations per second. Its field reliability will also be improved, pushing its "mean time between critical failures" beyond 2,000 hours. "We can gain a tenfold increase in radar sensitivity by replacing the computers," stated air force colonel James C. Bash in February 1988. "This allows us to spot a target ten times smaller than before at the same distance."

JOINING ARMS IN JOINT STARS

Beyond the navy's Aegis system and the air force's AWACS, few tactical C3 programs are confined to a single branch of the armed services. Interservice C3 programs have grown steadily since at least 1958, when President Dwight D. Eisenhower informed Congress that single-service operations were "gone forever." Although Eisenhower's pronouncement was premature—the services often resisted what they saw as dilutions of their identity—interservice C3 programs began to proliferate dramatically in the 1980s.

One combined venture, the Air Force-Army Joint Surveillance Target Attack Radar System (Joint STARS), may facilitate the management of warfare on the ground as much as AWACS has eased the management of war in the skies. Designed for use by NATO forces in Western Europe, Joint STARS would be crucial to the destruction of targets in the enemy's rear lines. The system is intended to enable the two services to pinpoint enemy armored and mechanized units en route to the battlefront rather than guess at their locations on the basis of dated surveillance reports.

Joint STARS features radar, computer, and communications technologies far more sophisticated than those originally incorporated in AWACS. The heart of the system is an EC-18 aircraft—like an AWACS plane, a converted Boeing 707—that embodies a phased-array radar, a computerized operations-and-control system run by two million lines of program code, and digital communications gear. The radar is carried in a canoe-shaped radome slung beneath the forward fuselage. Its signals are processed by a central computer and distributed among fifteen operations-and-control consoles on the plane, each able to display multicolored images of behind-the-lines terrain—including the vehicles moving across it. The central computer's massive memory enables console operators to view instant replays of enemy movements and to plot the target vehicles' courses and speeds. This in turn allows commanders of friendly air and artillery units to predict the sites at which the advancing enemy armor will be most vulnerable.

The army's portion of Joint STARS is its Ground Station Module, a van crammed with computers and communications equipment that receives data from the EC-18s and disseminates it to ground commanders.

COMBINING COMMUNICATIONS

Joint STARS aircraft and ground stations will be able to communicate with one another and with a host of air- and ground-combat units over a 345-mile range via JTIDS, the digital-data distribution system that would have been of such benefit in the Grenada invasion.

An operator monitors the movements of combat units on a display screen at the master station of a Position Location Reporting System. Using multilateration—an advanced form of radio direction finding—computers in the master station can simultaneously track hundreds of infantry units, vehicles, and aircraft dispersed across several hundred square miles. Accurate to within twenty-five yards, the system allows field commanders to direct maneuvers at night and in bad weather. A padlock secures equipment used to decrypt field transmissions, scrambled for security.

JTIDS is regarded as a breakthrough in preventing the saturation, or "self-jamming," of military communications channels. It uses a technique called time division multiple access (TDMA), in which a given cycle of transmission time—typically, about thirteen minutes—is split into 100,000 discrete segments, or slots. Each of the system's communicators—infantry units, tanks, planes, or even individual smart missiles—can then send a message through the system when the allotted time segment comes around. Units such as smart missiles that must report their progress faster than that can be accorded a transmission slot every few milliseconds or so. Messages consist of information transmitted in a strict sequence: identity, position, speed, mission, fuel and ordnance reserves, and so on. This time-sharing arrangement makes it possible to put an entire battle force—even one composed of air, sea, and land elements, as was the case in the Grenada operation—on a common, quicksilvery communications network that can update the status of 3,000 or more units twice a minute.

JTIDS will also make it difficult for a foe to jam or intercept messages. This electronic integrity stems from the use of two computer-controlled techniques, spread spectrum and frequency hopping. In conventional communications, the signals that make up each radio message are transmitted over a bandwidth just wide enough to accommodate them; in JTIDS, however, a computer spreads those signals over an extremely wide bandwidth, or spectrum, forcing an enemy who would jam them to likewise spread out—and in the process weaken—his interfering signals. Even when the jammer succeeds in disrupting a JTIDS message, the intended recipient can fall back on a clever piece of software code, known as a Forward Error Correction algorithm, that allows portions of a lost transmission to be reconstructed.

A second stratagem, called frequency hopping, aims to prevent the capture rather than disruption of radio signals. Frequency hopping involves the use of a computer to rapidly switch, or "hop," the signals among multiple radio frequencies at hundreds, even thousands, of hops per second. The computer orders the hops in a pseudorandom but actually predetermined sequence that is followed by all the radio sets—both transmitters and receivers—on a particular network. Such a ploy keeps would-be eavesdroppers guessing about which channel carries the target data.

When JTIDS terminals first entered service in the late 1970s, they were so big—about the size of home refrigerators—that they could be installed only on large platforms such as AWACS aircraft. The newest JTIDS terminals, by contrast, measure merely one-and-a-half cubic feet—about the size of a bread box, or small enough to fit in fighter aircraft and compact land vehicles. Meanwhile, JTIDS program officials plan to develop even smaller terminals, measuring just half a cubic foot, by fitting them with high-density, high-speed microchips called VHSICs (Very High Speed Integrated Circuits).

VHSIC chips, also slated for AWACS computers, may become the microelectronics hallmarks of virtually all C3 systems in the 1990s. Such chips were developed in the early 1980s in a billion-dollar program funded by the Department of defense, which foresaw that existing chips could not provide the computational power needed by emerging weapons (notably avionics-driven aircraft). The first VHSIC chips, produced in 1983 by a Cleveland-based electronics, aerospace, and defense company called TRW contained as few as 13,000 transistors and performed twenty-five million operations per second. TRW has since introduced a VHSIC chip embracing millions of transistors and capable of performing 100 million operations per second. Because malfunctions usually occur at the connections between individual chips, the increase in the number of transistors on a single chip will decrease the number of chips—and therefore the number of interchip connections—required in military computers. This in turn will greatly improve the computers' reliability.

THE SWEEP OF STRATEGIC C3

The computers that spin together the worldwide spider web of strategic command, control, and communications systems have a far more diverse and demanding mission than those responsible for tactical C3. They must interweave the communications network that connects the National Command Authority (the president, the secretary of defense, and their designated successors), the Joint Chiefs of Staff, and commanders-in-chief around the globe.

During a nuclear attack on the United States, that communications network would be set in motion by a message flashed from early-warning satellites, airborne radars, or land-based radars to NORAD headquarters, deep inside Cheyenne Mountain, Colorado. NORAD computers would then forward the harrowing news to the Strategic Air Command at Offutt Air Force Base in Nebraska, to the National Military Command Center at the Pentagon, and to the underground Alternate National Military Command Center at Fort Ritchie, Maryland, in the Catoctin Mountain range. Time permitting, the National Command Authority (NCA) and the Joint Chiefs of Staff would also take to the air in the National Emergency Airborne Command Post (NEACP, pronounced "knee-cap") aircraft. SAC's command-post aircraft and the navy's various TACAMO (Take Charge and Move Out) aircraft for communicating with its ballistic-missile submarines at sea would already be aloft, since they are on airborne station around the clock.

Such emergency measures constitute the front end of a U.S. strategic command, control, and communications system that, as Assistant Secretary of Defense for C3 Donald Latham noted in 1984, "must be perceived to be convincingly capable of responding to enemy aggression."

Binding the whole affair is the Worldwide Military Command and Control System (WWMCCS, pronounced "wimmex"), a network of twenty-six command centers using thirty-five central computers programmed with a total of more than twenty million lines of code. The main computers are linked with one another via the WWMCCS Intercomputer Network (WIN). All WWMCCS computers and communications facilities are also tied into those of the Defense Communications System (DCS), a much larger network of digital voice and data communications that links major U.S. commanders everywhere.

Computers form the soul of WWMCCS. Located in strategic command-and-control centers worldwide, the computers are designed to allow the National Command Authority and top commanders at the Pentagon to monitor all U.S. military operations around the globe, those tactical as well as strategic. In its bid to fill this role, WWMCCS initially displayed some unnerving shortcomings. For example, during a 1977 test of WWMCCS' ability to link the command-center computers of all the services in the United States, Europe, and the Atlantic, the sample emergency-action messages—coded strings of fewer than 250 bits, conveying data about the target selected and the time for the attack—got through only 38 percent of the time. Since then, WWMCCS' performance has been dramatically improved, with test messages passing successfully through the system 95 to 99 percent of the time.

TIGHTENING THE WWMCCS WEB

The growing pains that WWMCCS has undergone stem partly from the Pentagon's efforts to keep the system's computer technology up to date. As early as 1971, in fact, the Pentagon had launched a program to connect the computers in the system through the medium of a single intercomputer network. Network-control programs were introduced that allowed one computer to send messages to all others over common communications channels.

This setup has since grown much more sophisticated and secure. It now features a technology called packet switching, which DARPA originally developed in the 1970s for a computer network linking the Defense Department with universities involved in its research projects. Packet switching involves dividing data streams, or messages, into packets of 1,000 bits each. This is performed by a computer called an Interface Message Processor, or IMP, that is connected to the message-sending computer. The IMP also encrypts the packets, then sends them piecemeal and out of order to a receiving computer, which hands them over to its own IMP for decoding and reassembly into the original message. If an eavesdropper managed to intercept a message, it would be so jumbled that deciphering the contents would present monumental difficulties.

Until the 1980s, the Pentagon's modernization of WWMCCS was distinguished by incremental improvements in software and hardware. But in 1982, as part of the Reagan administration's strategic modernization program, the

Pentagon set out to replace much of the WWMCCS software and all of its mainframe and work-station computers with machines of much greater speed, power, and reliability. Dubbed WIS, for WWMCCS Information System, the endeavor will stretch well into the 1990s.

WIS boasts a flexible "total system architecture"—that is, its overall makeup has been designed to accommodate different families of mainframe computers in order to satisfy the special computing and communications requirements of each WWMCCS command center. Such flexibility was not afforded by the original WWMCCS architecture, in which all command centers, regardless of their individual mission, were equipped with Honeywell 6000 computers, first manufactured in 1964.

A second important feature of WIS architecture is distributed processing. Each command center will be furnished with a local area network (LAN) designed to interconnect its internal computers—mainframes, work stations, and mass-storage devices—with one another and with another computer, called a network interface processor, that channels message traffic into and out of the command center itself. Distributed processing will allow the computer system of each command center to be upgraded in modular fashion, enabling WIS to keep pace with advances in computer technology. It will also allow the widespread dispersion of personnel and computing facilities in mobile command centers, typically aircraft or transportable military shelters. Such distributed command centers will improve the system's survivability by increasing the number of targets an enemy must hit in order to disrupt communications.

THE ESPERANTO OF SOFTWARE LANGUAGES

Distributed processing like that adopted for WIS would be impossible without a common software language that allows computers of various architectures to converse with one another. The software chosen for WIS is therefore written in a language called Ada. Developed under Pentagon auspices between 1975 and 1979, Ada was formally adopted by the Defense Department in 1983 for all "mission-critical" software—programs that govern the operation of combat systems, among them Aegis and AWACS.

Ada is modular in nature. It allows programs to be written in self-contained packages—each designed, tested, and debugged by a separate programmer or programming team—that are then fitted together into a whole. This building-block approach yields easily maintained, easily understood programs. Ada's special appeal, however, derives from having been copyrighted by the Defense Department. Owning the language has permitted the Pentagon to insist on rigorous standardization, with the result that Ada exists in only one version. Other, less tightly controlled languages have been customized or expanded by several different publishers, and the multiple versions are not fully compatible with one another. The absence of a language barrier within Ada helps to ensure that sections of software written by one contractor will mesh with programs written by another.

So great are the advantages of programming in Ada that by the late 1980s the language had been incorporated in or readied for more than 130 military systems, ranging from army battlefield computers to the air force's $40 million Advanced Tactical Fighter. Although the military continues to use more than

A technician aligns an antenna of a mobile radio station capable of picking up battlefield communications transmitted from beyond the horizon. Such communication is made possible by the tendency of high-frequency radio waves to be scattered by turbulence in the lower atmosphere. A pair of antennae captures the weakened signals, and signal-processing equipment eliminates distortion to produce an intelligible message. Such an outpost might relay a warning of approaching enemy aircraft from a phased-array radar installation to a commander of ground forces.

forty programming languages for a broad range of applications, the widespread acceptance of Ada will encourage that babble to subside over time.

COMPUTER INSECURITY

Although Ada was designed to let war-fighting computers speak in a common tongue, the need for military machines to exchange their information—and thereby expose it to the risk of electronic capture—forms an enduring chink in the Pentagon's armor. Safeguarding computer data has traditionally hinged on how the machines are operated or accessed, not on how they are designed or constructed. The information in command-center computers, for example, can usually be tapped only by personnel with security clearances for the information. The most sensitive data is processed in separate computers that are physically and electronically isolated from the others; it may also be processed by authorized operators during select intervals, with the data files being sequestered at the close of each session. Yet such methods are costly, awkward, and inefficient, and in wartime they would likely go by the boards.

In response, computer designers began building security into the very makeup, or architecture, of computers and communications systems. To adapt such design features for its own use, the Department of Defense established the Computer Security Evaluation Center (CSEC) in the early 1980s. The CSEC certifies the level of security that any given military computer meets.

Computers can also be compromised by the capture of the electromagnetic emissions, or "noise," that the machines' circuitry and cables radiate into their surroundings. By intercepting such seeping radiation, no matter how minuscule it may be, eavesdroppers can often use their own computers to figure out the full

Self-Defense for Military Computers

In military applications, computers can be surprisingly vulnerable. For instance, a computer's central processing unit, magnetic storage devices, and communications gear all emit electromagnetic waves, like a radio transmitter broadcasting the computer's business. Without safeguards, data can be intercepted by eavesdropping devices up to half a mile away.

In addition, wartime conditions can damage electronic circuitry. The greatest threat is an electromagnetic pulse

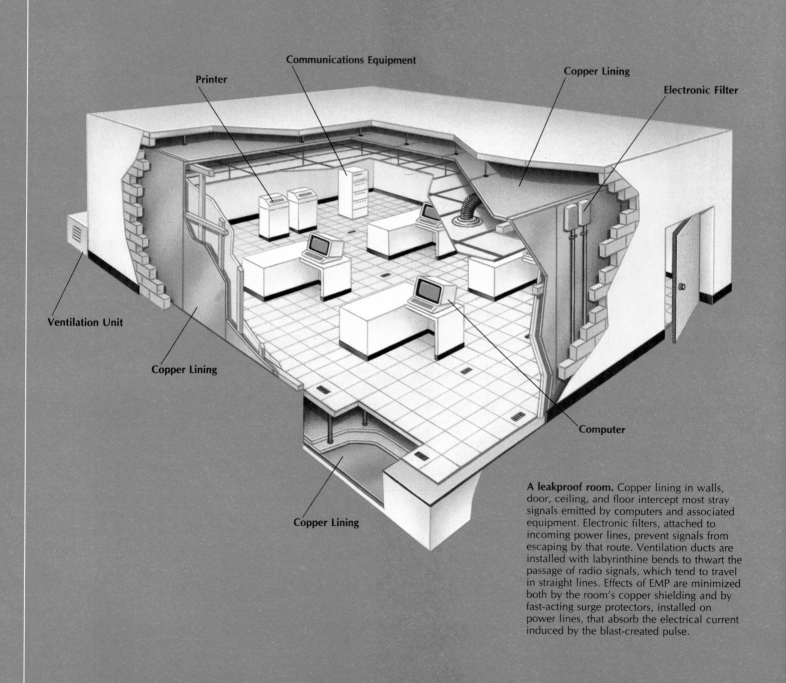

Printer

Communications Equipment

Copper Lining

Electronic Filter

Ventilation Unit

Copper Lining

Copper Lining

Computer

A leakproof room. Copper lining in walls, door, ceiling, and floor intercept most stray signals emitted by computers and associated equipment. Electronic filters, attached to incoming power lines, prevent signals from escaping by that route. Ventilation ducts are installed with labyrinthine bends to thwart the passage of radio signals, which tend to travel in straight lines. Effects of EMP are minimized both by the room's copper shielding and by fast-acting surge protectors, installed on power lines, that absorb the electrical current induced by the blast-created pulse.

(EMP), a burst of intense magnetic energy from a nuclear blast. The EMP produced by even a distant explosion can generate an electrical current strong enough to melt circuitry inside computers and peripheral equipment. But protection is afforded by the simple tactic of enclosing the equipment in metal; the electromagnetic pulse generates a current in the casing, where it can be drained away safely. Computers may be shielded individually, but they are often protected in groups by safeguarding an entire room or a building (below, left).

Computers that operate on the battlefield, whether built into a tank or an aerial-reconnaissance drone, also face environmental stresses sufficient to incapacitate any office computer. Such computers (below) are therefore designed inside and out to withstand the hard knocks of combat, dust, smoke, fungus, humidity, and salt spray, and to operate faultlessly at temperatures that may range from -60° F. to 160° F.

This ninety-pound battlefield computer is about two feet long and eight inches tall. The front panel includes the switch to turn the device on and off, as well as connectors for a power source and a range of input/output devices.

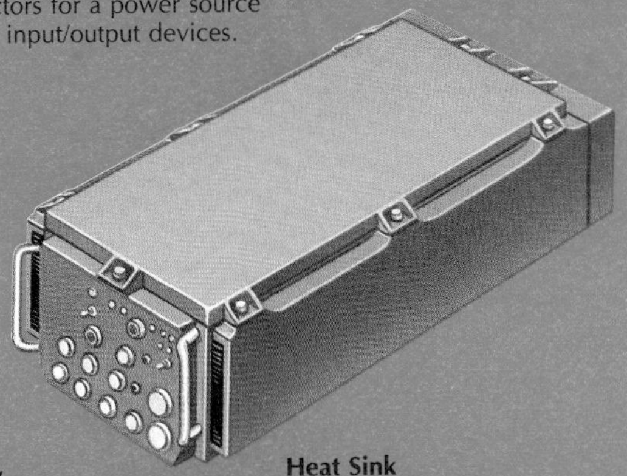

Inside a battlefield computer. Aluminum stiffeners, a rigid chassis, and a tough outer case help this computer, shown in cutaway view, withstand the abuses of combat. A layer of copper on each circuit board serves as a heat sink, conducting heat (red arrows) from the chips to the chassis. Another heat sink extracts heat from the power supply. Cool, filtered air (blue arrow), circulated by a fan, absorbs the heat and expels it through vents in the front panel.

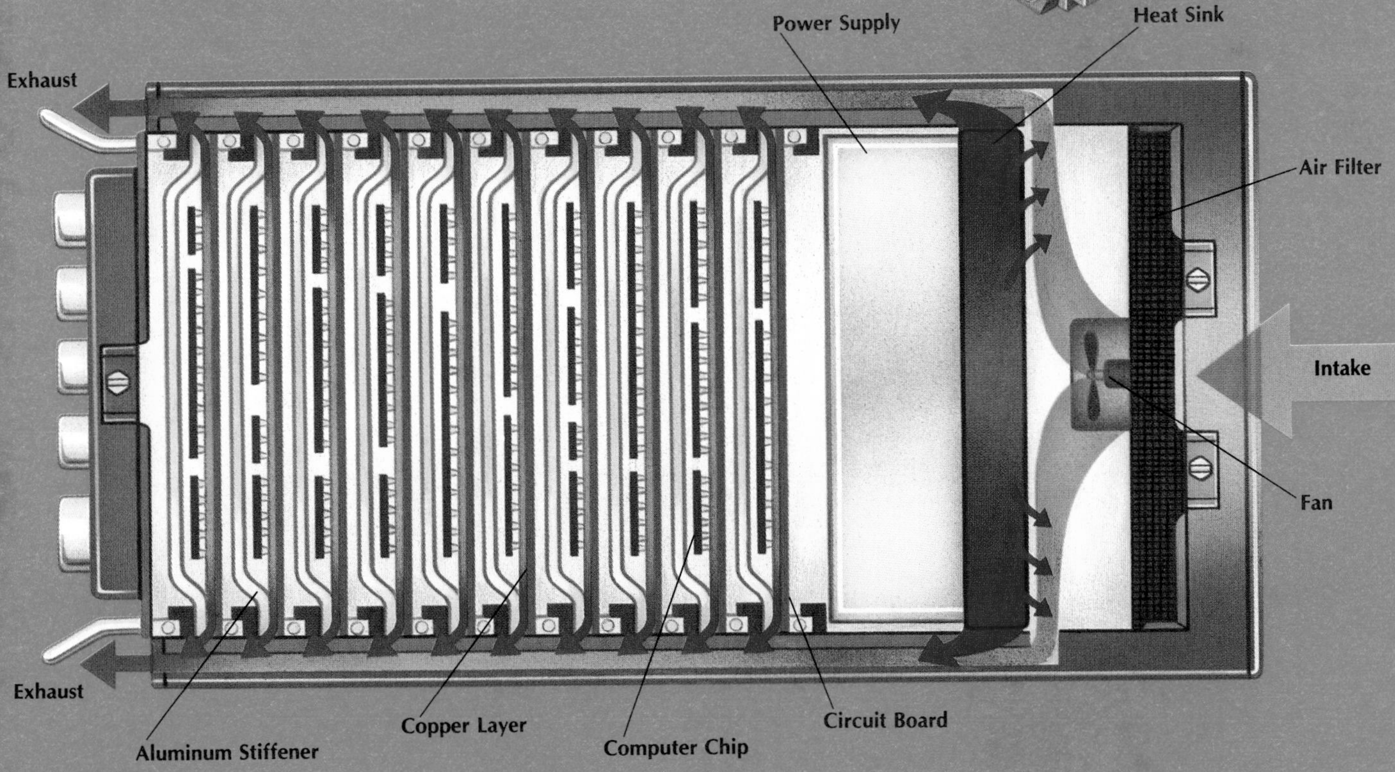

Power Supply

Heat Sink

Exhaust

Air Filter

Intake

Fan

Exhaust

Aluminum Stiffener

Copper Layer

Computer Chip

Circuit Board

text of the information the target computers are processing. To bottle up these telltale signals, the Pentagon has encased entire computer rooms in metal *(pages 104-105)*. The wires carrying classified data back and forth among the computers must likewise be protected. In some cases, this is achieved by isolating the wires in gas-filled tubes; any attempt to tap the lines causes the pressure in the tubes to drop, alerting the communicators that their network has been infiltrated.

Future security problems may be solved by photonic computers—machines operated by light beams, made up of photons, rather than by electrons. Lacking electronics, photonic computers would be unjammable. They would also be more resistant to radiation, notably the powerful electromagnetic pulse (EMP) that results from a nuclear explosion.

THE PULSE OF AN ATTACK

In 1984, a Pentagon official characterized EMP as a "cheap shot" that could "decapitate the National Command Authority from the force commanders so that we couldn't launch the force." Indeed, by its indiscriminate capacity to cripple all manner of C3 systems, EMP is a phenomenon of frightening proportions. "In the event of heavy EMP radiation," wrote nuclear physicist Edward Teller in 1982, "I suspect it would be easier to enumerate the apparatus that would continue to function than the apparatus that would stop."

An electromagnetic pulse takes place when the gamma rays given off by a nuclear burst interact with the air, producing a transitory but immensely powerful energy wave that induces a voltage in the conductors of every electronic device in its path. A high-altitude nuclear explosion would short-circuit much of the electronic apparatus of everyday life: Electronic watches would freeze like stop clocks, television sets would go black, and cars sparked by electronic ignition systems would not respond to the turn of the key.

For a computer exposed to EMP, the least harmful effect would be the loss of all data resident in its electronic memory. Depending on how much energy the EMP brought to bear, the computer's semiconductors and their electrical connections might be permanently disabled. This mortal blow would be inflicted with almost unimaginable rapidity; a typical EMP lasts only about 200 nanoseconds, or billionths of a second. A lightning bolt, by contrast, builds to peak strength in a period 10,000 times longer than that.

The wide-ranging effect of EMP was first observed on July 9, 1962, when the United States detonated a 1.4-megaton hydrogen bomb 250 miles above Johnston Island in the Pacific Ocean. The electromagnetic pulse from that blast tripped burglar alarms and shut off lights in Honolulu, some 800 miles away. The irony is that EMP is one million times more destructive of today's exquisitely miniaturized solid-state integrated circuits (ICs) than it is of the vacuum tubes that ICs replaced years ago in computers.

By the mid-1980s, radiation shielding was being provided for the computers at the heart of U.S. command centers and other C3 facilities, as well as for the computers that help control strategic weapons. Such shielding can take a variety of forms. Critical circuits may be sealed in a conductive shell (good conductors such as gold, silver, and copper are best); the electrical energy of the pulse travels around the shell and is bled away from the circuits inside. Alternatively, the circuits may be fitted with surge arresters—electronic valves designed to stanch current-flow whenever they detect a sudden increase of voltage. Where it is impossible to protect ICs made of silicon—a material that is relatively susceptible to EMP—circuits made of the EMP-resistant compound gallium arsenide may be used instead.

Given the cost and scope of the task, however, there seems to be little or no hope of EMP-proofing everything in strategic sight. Testing conducted by the air force in 1985, for example, confirmed designers' suspicions that it would be prohibitive in terms of expense and weight to thoroughly shield from EMP such electronics-packed aircraft as the F-15 and F-16 fighter and the B-1B bomber.

STAR WARS: A TITANIC TASK

Strategic command, control, and communications assumed herculean responsibilities as the 1980s progressed. C3 was called upon to form the basis of a ballistic-missile defense system, postulated by President Reagan in his so-called "Star Wars" speech of March 23, 1983. Reagan envisioned a space-based shield so effective that it would render nuclear weapons "impotent and obsolete." The proposed system was thereafter nurtured, amid much debate over its feasibility, in the administration's Strategic Defense Initiative (SDI)—a program to develop a multilayered defense against ballistic missiles in all four phases of their flight. These four phases are the boost phase, the powered portion of a missile's flight; the post-boost phase, during which warheads and decoys separate from the final stage of the missile; the midcourse phase, during which the warheads and decoys follow a ballistic trajectory through space toward their targets; and the terminal phase, when the warheads reenter the earth's atmosphere and descend upon their targets. Whatever weapons are adopted for downing the missiles and their warheads, computers will be crucial to arm, aim, fire, guide, and coordinate them.

By mid-1987, the Pentagon's Strategic Defense Initiative Organization (SDIO) had concluded that steerable projectiles, also known as "smart rocks," could be deployed in space and also on land to hit and destroy ICBM boosters, their warhead-carrying buses, and the warheads themselves. Tests suggested, however, that such weapons could not be fielded until the late 1990s.

According to the SDIO's conception of these kinetic-energy weapons (KEWs), several hundred weapons platforms in low orbit would carry ten missiles apiece to intercept ICBM boosters before their buses could release any warheads. In space, the KEWs would be teamed with surveillance and tracking sensors on the lookout for ICBMs in their boost and post-boost phases. Additional KEWs would be positioned on the ground and would work in concert with those in space to intercept reentry vehicles from the time they left their buses in the post-boost and midcourse phases to the time they penetrated the lower atmosphere in the terminal phase.

The SDIO also began investigating the possibility of deploying directed-energy weapons (DEWs) as futuristic complements to the KEWs. Research on DEWs— high-powered laser beams that would burn through the metal exterior of ICBMs and detonate their fuel, and particle beams made up of neutral atoms such as hydrogen that would savage the missiles' electronics—showed they could not be deployed any time soon. For the near term, however, the particle beams might serve as devices that would aid in discriminating real targets from the fake ones that ICBM buses disgorge in space. These spurious targets include decoys—

typically, balloons made of radar-reflective aluminized polyester film—and radar-baffling chaff, composed of streams of thin wires that surround and camouflage the live warheads.

WHEN COMPUTERS CALL THE SHOTS

KEWs and DEWs will be worthless, however, unless they can be orchestrated by a C3 system of unparalleled performance. SDI's goal is to be able to thwart an all-out ICBM assault by the Soviet Union, which would throw at the United States an estimated 3,000 long-range ballistic missiles carrying as many as 25,000 warheads and 250,000 to 300,000 decoys. Success by the defense system would require unjammable communications operating at extraordinarily high data rates in weapons, in weapons platforms, in surveillance satellites, and in C3 centers both in satellites and on the ground.

According to the battle scenario pictured by the SDIO, infrared sensors aboard early-warning surveillance satellites would detect and track enemy ICBMs as they vaulted spaceward. Those satellites would then alert human battle managers at control stations on the ground, who would send a "go-code" authorizing the use of weapons to computers aboard the orbiting interceptor platforms. As the

The second stage of a Titan I missile disintegrates in the beam of a laser during a 1985 test at White Sands Missile Range, New Mexico. A two-second exposure to the two-megawatt laser, fired from about a mile away, caused the booster to burst under the simulated stress of flight. For a laser to be practical as a weapon against ICBMs, it would have to destroy the rocket in a fraction of a second at a range of 2,000 miles or more, a feat that would require a beam more than 100,000 times as powerful as this one.

ICBMs came within range of each platform, the platform's processors would assign interceptor missiles to their targets and launch them. Kill-assessment computers on the platforms, meanwhile, would keep track of which warheads had been destroyed and which had leaked through to the next layer of the defense shield. Such a degree of computerization, concluded air force general Melvin Chubb in 1985, "pushes battle management as far as you can push it."

Although human battle managers would oversee the entire process—an arrangement known as keeping "the man in the loop"—some critics of SDI contended that those managers would be almost ancillary to the direction of the actual combat. Computers would select targets, assign weapons, and control the engagements. Such a defense would therefore "have to be almost automatic," says C3 expert Don Latham. "If you're going to do a boost-phase kill, it's got to be activated within two to three minutes after the boosters leave the ground." To be sure an attack was under way, Latham adds, "some initial leakage in the boost phase would be acceptable. The first ten or twenty or 100 ICBMs would get engaged later so that the boost-phase system is not trigger-happy."

SDI SOFTWARE: EQUAL TO THE TASK?

Controversy will no doubt continue to embroil the SDI, yet on one point its proponents and detractors have long agreed: Writing the programs to run the system's computers will be the greatest software challenge of all time.

Whether or not the U.S. computer industry can design that software became a topic of intense speculation soon after the Strategic Defense Initiative program got under way in late 1983. By October of 1985, when 1,200 computer experts crowded into the Kresge Auditorium of the Massachusetts Institute of Technology to witness a debate on the topic, the lines of argument had already been clearly drawn. Those supporting the feasibility of SDI software were represented by computer-science professor Danny Cohen, who stated that the system's computers need not be driven by a single, monolithic program. Instead, Cohen suggested, the software could be produced in modules, each controlling a separate family of sensors and weapons; with programming glitches confined to individual modules, the task of isolating and correcting them would be simpler.

Those questioning the reliability of SDI software were represented by computer scientist David Parnas, who stressed the impracticality of testing the programs, no matter what form they took. "It might not be impossible to put something up there that will work," admitted Parnas, "but there will never be a day when we will trust it."

How large the software for SDI might be—and how best it could be put to use—became subjects of equally animated conjecture. In April 1987, SDIO director General James Abrahamson estimated that the SDI system "will have to have more than ten million lines of code." But most software experts predicted that the total programming needed would range from forty million to 100 million lines. General Abrahamson also forecast that the software's "very effective architecture" would make it conducive to being implemented as a highly distributed system that "doesn't have to come together in one supercomputer."

The software for such a distributed system would lend itself to testing before deployment in the form of computer simulations. Such dry runs will be conducted by Cray supercomputers at Martin Marietta's National Test Bed facility outside Colorado Springs, Colorado. The supercomputers there can be programmed to simulate various missile-attack scenarios and gauge the most effective ground- and space-based defenses against them.

The software for an antimissile defense system might be much simplified by adopting a smaller version of the smart rock called a brilliant pebble. First proposed by researchers at the Lawrence Livermore National Laboratory in 1986, these smaller projectiles, each with its own targeting computer, could be sent into orbit individually and programmed to locate targets on their own, without comprehensive battle management software. During a nuclear attack, ground controllers would simply command the tens of thousands of orbiting projectiles to proceed autonomously with the destruction of incoming ICBMs. With so many brilliant pebbles aloft, it would matter little if more than one attacked a single warhead.

HARDWARE FOR THE 1990s

While the SDIO grapples with the prospect of creating a weapons system that must work reliably the first and only time it is used, DARPA—the Pentagon's blue-sky research-and-development arm—continues to spur the development of new computer hardware of all kinds. As part of the Strategic Computing Program that it inaugurated in 1985, DARPA is underwriting the production of prototype supercomputers that will use parallel processing and gallium arsenide chips to crunch numbers 100 times faster than the swiftest existing mainframe. Meanwhile, DARPA's projections for quantum leaps in speed have been rivaled by startling breakthroughs in the miniaturization of components. At the air force's Rome, New York, Air Development Center (RADC) in 1987, for example, engineers began assembling a supercomputer that consisted of silicon wafers stacked one atop the other in a container the size of a three-pound coffee can. When it is completed in the mid-1990s, the supercomputer will be able to execute between 100 billion and one trillion operations per second while consuming just thirty watts of electrical power. That performance would fulfill DARPA's prophecy of a hundredfold increase in computing speed; at the same time, it would be achieved with one hundredth the amount of current required by contemporary supercomputers.

Miniaturized machines like the wafer-stack supercomputer will enable computation to migrate from the ground up to space, thwarting an enemy's attempts to jam the satellite communications that would set an antimissile defense in motion. Such communications presently invite disruption simply because they are so voluminous; today's early-warning satellites send their data to computers on the ground in a continuous, largely unprocessed stream. This uninterrupted flow of information would help the enemy pinpoint and jam the source of the transmissions. If the data could be processed by small, reliable computers onboard the satellites, messages could be flashed as high-speed data bursts among the sensors, computers, and weapons that compose an antimissile network. The system would be far less vulnerable to communications interference.

Perhaps such computers will wind up as indispensable elements of a space-based strategic-defense system that provides a powerful deterrent to nuclear attack, or that limits the damage should the unthinkable occur. If so, computers will have fulfilled a transcendent military role—as weapons against war itself.

A Shield
in Space

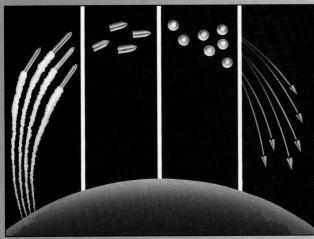

Boost   Post-Boost   Midcourse   Terminal

A full-blown nuclear-missile attack is the ultimate nightmare of modern war, and defending against it represents the ultimate technological challenge. While not yet proved feasible, a space-based defense against intercontinental ballistic missiles (ICBMs) would doubtless lean heavily on computers to handle the enormous volumes of data that would be involved in detecting, tracking, and intercepting thousands of individual targets.

The ability to act quickly and adapt readily to changing circumstances would be indispensable to any such defensive scheme. ICBMs would take about half an hour to reach their targets, traveling in a high arc that subsumes four distinct phases of flight, as illustrated at left.

In the boost phase, lasting for only about three to five minutes, missiles would be propelled up through the atmosphere by massive booster rockets. In the post-boost phase, lasting another three to five minutes, missiles separated from their boosters would prepare to release as many as ten warheads apiece, as well as hundreds of decoys.

The midcourse phase would last for another twenty minutes, as the warheads and decoys sail thousands of miles through space in ballistic trajectories; encased in shiny metallic balloons, the warheads would be all but indistinguishable from empty balloon decoys. In the brief terminal phase, lasting less than a minute, the decoys would disintegrate or slow as they reentered the atmosphere, and the warheads would speed toward their targets.

During each of these phases, specially tailored sensors and weapons would work to counter the threat. Highly sophisticated devices are envisioned for the task, but the most critical and intricate equipment of all would be the computers charged with managing the system.

Orchestrating a Response

One of the most complex features of a space-based defense system would be battle management—coordinating the actions of perhaps thousands of defensive components in order to destroy as many missiles and warheads as possible. The task would be the responsibility of an elaborate network of computers working together to handle, in the few short minutes available, a flood of details that would quickly overwhelm human controllers.

One likely strategy would be to distribute computer control to a fleet of battle-management satellites, each responsible for a particular group of sensors and weapons. These local battle managers would analyze reports from their sensors and create separate tracking files for each missile, warhead, or decoy in their area. The managers would then assign targets to the various weapons systems under their command and recheck sensors to determine if the targets had been successfully intercepted or if further shots needed to be fired. As the battle proceeded, they would pass their tracking files to the battle managers for the next phase of the defense so that targets managing to escape one round of interceptors could be picked up by the next.

In overall charge would be a global battle-management computer, stationed either on the ground or, as illustrated below, in a special airborne command post. It would correlate information from all the local battle managers to keep track of the course of events. Among its most important chores would be reconfiguring the defense as individual elements fell prey to the enemy's expected antisatellite attacks.

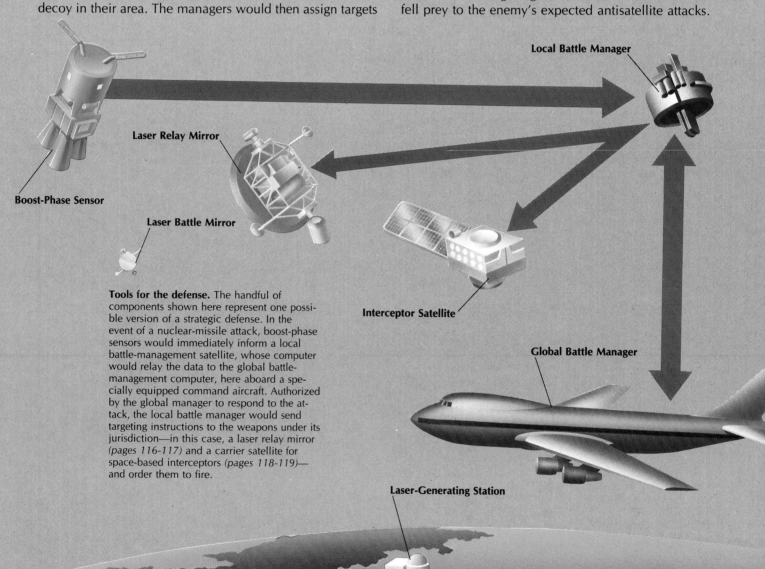

Boost-Phase Sensor

Laser Relay Mirror

Laser Battle Mirror

Local Battle Manager

Interceptor Satellite

Global Battle Manager

Laser-Generating Station

Tools for the defense. The handful of components shown here represent one possible version of a strategic defense. In the event of a nuclear-missile attack, boost-phase sensors would immediately inform a local battle-management satellite, whose computer would relay the data to the global battle-management computer, here aboard a specially equipped command aircraft. Authorized by the global manager to respond to the attack, the local battle manager would send targeting instructions to the weapons under its jurisdiction—in this case, a laser relay mirror (pages 116-117) and a carrier satellite for space-based interceptors (pages 118-119)—and order them to fire.

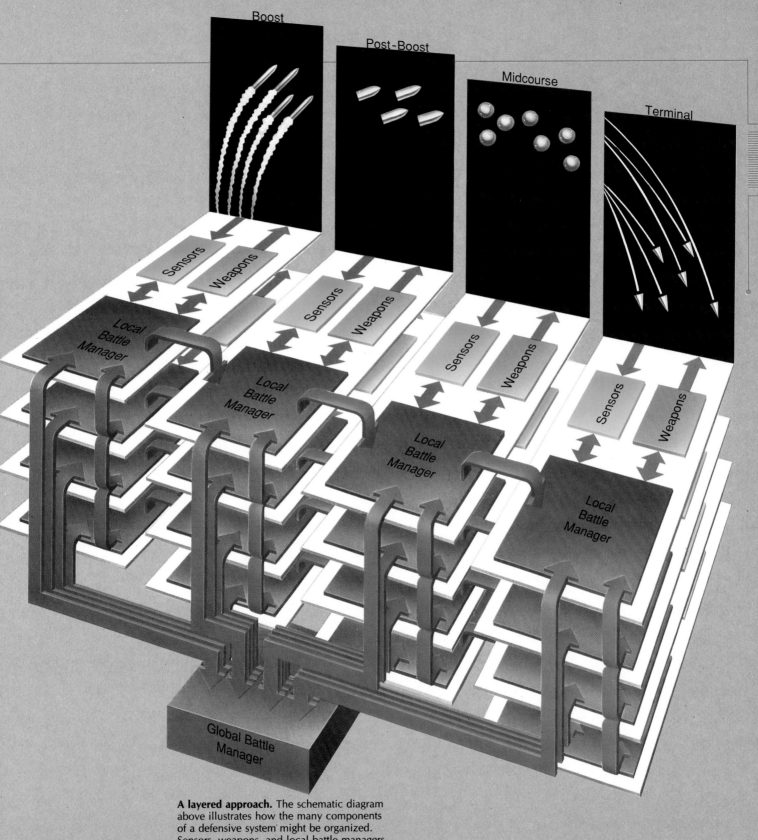

A layered approach. The schematic diagram above illustrates how the many components of a defensive system might be organized. Sensors, weapons, and local battle managers are divided into four groups, corresponding to the four phases of a missile attack; several levels exist within each group to serve as backups in case elements of one level are damaged or destroyed. Information and instructions flow between sensors and weapons and their local managers, which in turn communicate with the global manager, as well as with their backups and local managers in the next phase.

A Shield in Space

Boost-phase detection. An early-warning satellite is positioned in orbit so that its infrared sensors, protected from peripheral radiation by sun shields, can detect the superhot exhaust plumes of ascending enemy missiles. By analyzing plume images *(box, below)*, onboard microprocessors quickly determine how many missiles were fired and where they are headed, then forward the information to a local battle-management satellite. The early-warning satellite can continue tracking the missiles or shift its gaze toward other possible launch sites.

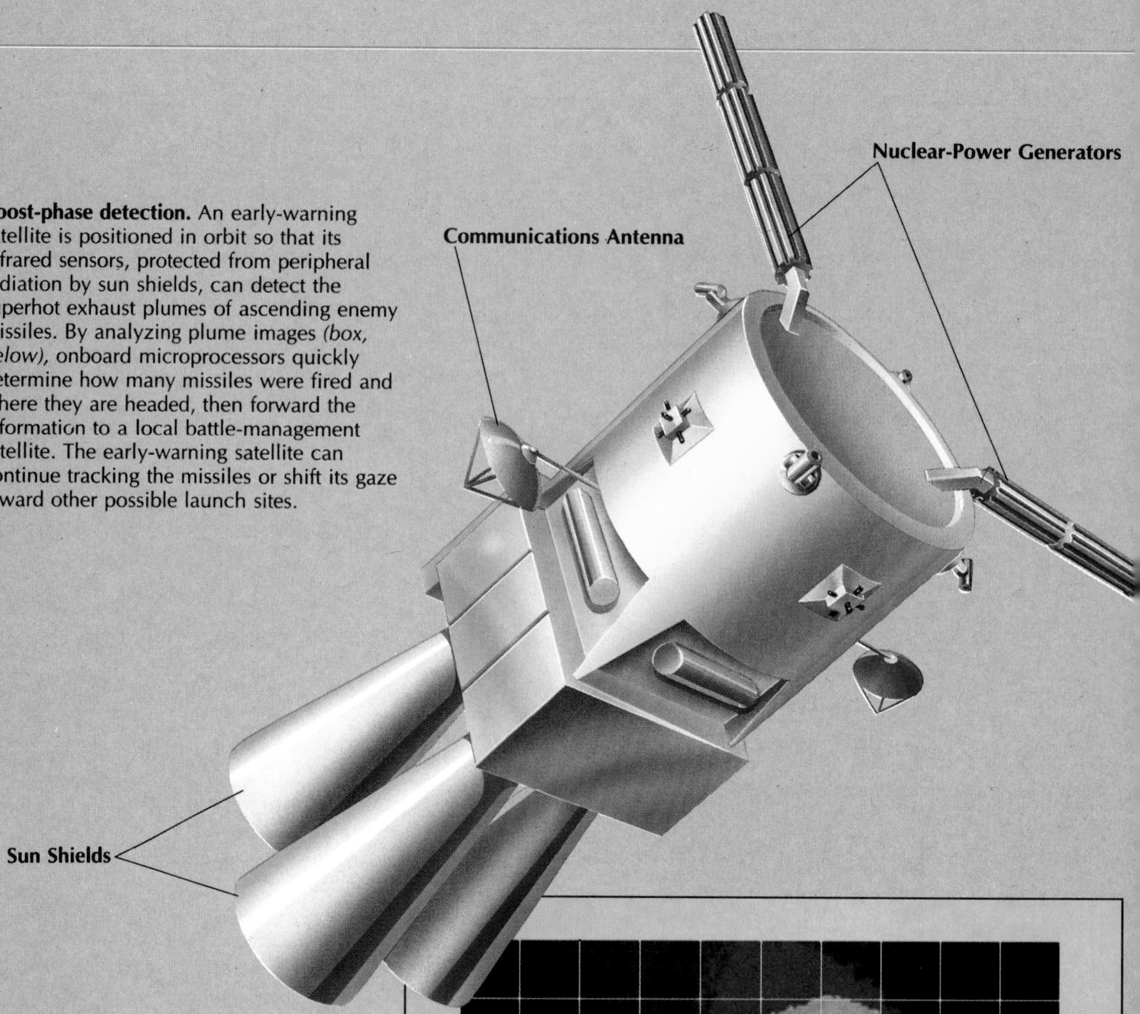

Nuclear-Power Generators

Communications Antenna

Sun Shields

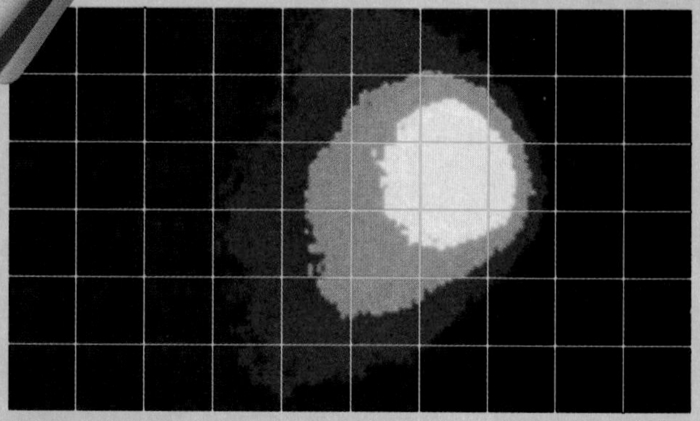

An electron-based image. Infrared radiation from a rocket plume is detected by a grid of sensors known as a charge-coupled device (CCD). As infrared light strikes the CCD's surface, it excites electrons at each picture element, or pixel, of the grid in proportion to the light's intensity there (for clarity, a sixty-pixel grid is shown above; the actual image contains hundreds of thousands). The charge at each pixel is measured, creating a digitized version of the image. Microprocessors analyze new images every few milliseconds, watching for the characteristic plume patterns of ICBMs and tracking their trajectories. False colors highlight this plume's temperature pattern, with yellow denoting the hottest part of the exhaust near the engines.

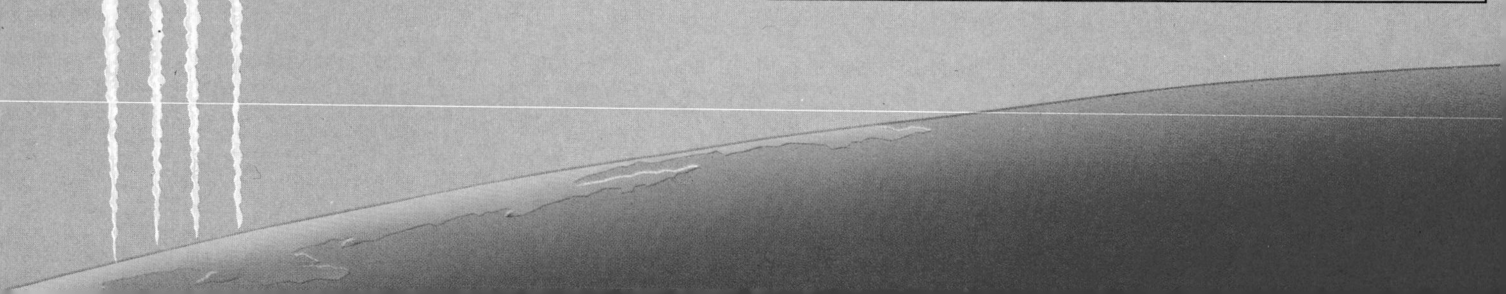

High-Tech Watchers in the Skies

Several types of sensors are envisioned as sentinels for a space-based defense, each specially designed to cover a different phase of missile flight. They fall into two broad categories, represented by the two satellites on these pages. Passive sensors *(left)* would emit no signals but rather detect radiation—such as heat—emanating from missiles and warheads; active sensors *(below)* would project sensing beams, then watch for reflections. In either case, a key component would be onboard microprocessors programmed to recognize patterns in detected signals, track objects as they proceed, and continually relay their findings to local battle managers.

At the very onset of an attack, passive infrared detectors such as the one at left would register the intense heat generated by the rocket engines that boost missiles up through the atmosphere. Once the engines shut down, active sensors would take over, illuminating the smaller, colder post-boost vehicles with radar signals or, perhaps, with low-energy laser beams. During the midcourse phase, when warheads within aluminized balloons look exactly the same as empty decoys, both active and passive sensing might be employed. Particle-beam accelerators such as the one below could spot warheads by penetrating their balloons with a beam of high-velocity atomic particles. And passive infrared detectors more finely tuned than their boost-phase colleagues might be able to pick out the warmer warheads from the cooler decoys around them.

For the terminal phase, ground-based detectors would come into play. The steerable beams of phased-array radars *(pages 28-29)* would provide crucial last-minute tracking of any incoming warheads that might have managed to escape the space-based defenses.

Seeing Through Decoys

During the midcourse phase, a particle-beam accelerator *(right)* distinguishes between empty decoy balloons and those containing warheads by firing a stream of hydrogen atoms at them. A beam-steering sensor locates the balloons, and the beam is aimed at each one in rapid succession. When a decoy is struck, the beam encounters little mass and passes right through. But when the beam hits a warhead, sparks of subatomic particles are emitted and are detected by another sensor on the accelerator. Microprocessors record which balloons contain warheads so they can be targeted for attack.

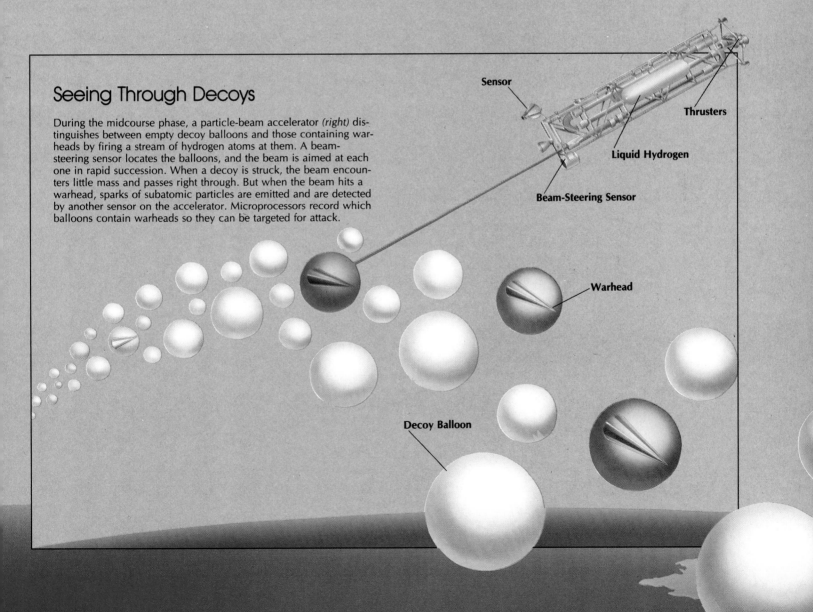

Sensor

Thrusters

Liquid Hydrogen

Beam-Steering Sensor

Warhead

Decoy Balloon

Killer Beams to Thwart an Attack

To be effective, a defense against nuclear missiles would have to be able to strike quickly and early, when warheads are still in their launch vehicles and several could be destroyed with one shot. Because of their potential ability to deliver lethal blows literally at the speed of light, laser weapons would undoubtedly be crucial to any such defensive scheme.

Lasers project light waves in a tight beam that does not disperse and weaken like ordinary light; as a result, they can focus a great deal of energy on a very small area. By concentrating massive amounts of radiation, a laser weapon could, for example, burn a hole through a booster rocket's fuel tank and cause it to explode. But to have access to the huge supplies of power it would need, such a device would almost certainly have to be based on the ground, perhaps thousands of miles from its targets—a requirement that would vastly complicate the weapon system and necessitate special computing support.

The simplified system illustrated here includes two space-based beam-directing mirrors measuring about fifteen feet across. To begin with, a computer at the beam-generating station on the ground would take steps (box, below right) to ensure that the beam reached the first of these mirrors without losing any of its potency on its way through the atmosphere. Then, based on sensor input about target locations, battle-management computers would speed through complex calculations to coordinate the orbital positions and reflection angles of the mirrors. Finally, microprocessors aboard the mirrors would execute the battle-management commands, making adjustments in split seconds in order to intercept as many targets as possible with the beam's deadly force.

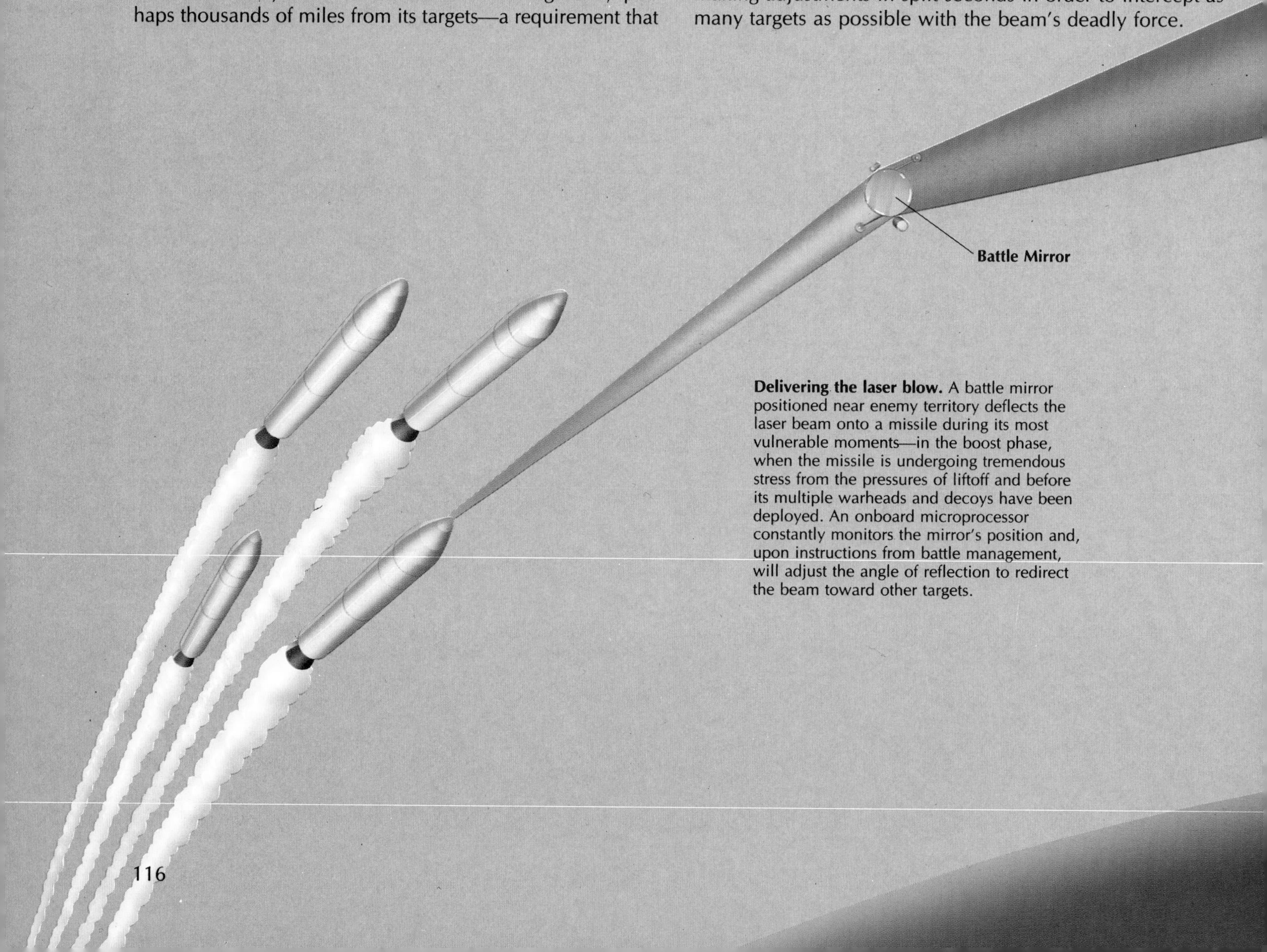

Battle Mirror

Delivering the laser blow. A battle mirror positioned near enemy territory deflects the laser beam onto a missile during its most vulnerable moments—in the boost phase, when the missile is undergoing tremendous stress from the pressures of liftoff and before its multiple warheads and decoys have been deployed. An onboard microprocessor constantly monitors the mirror's position and, upon instructions from battle management, will adjust the angle of reflection to redirect the beam toward other targets.

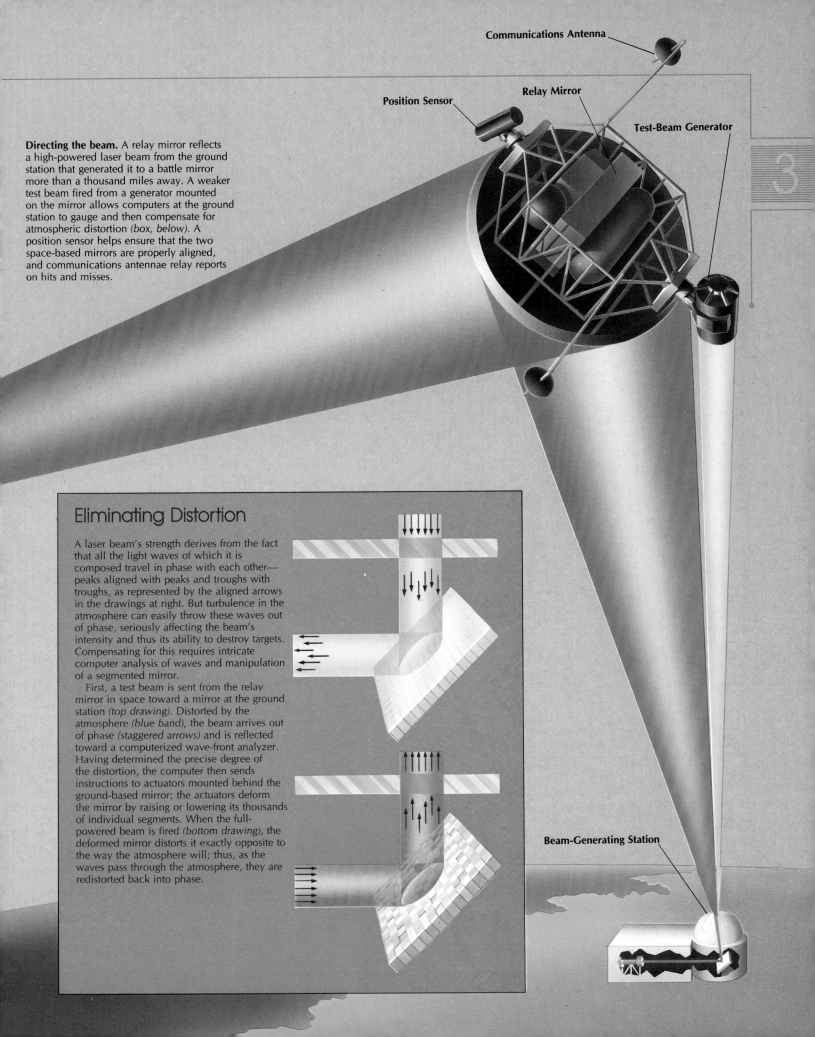

Communications Antenna

Position Sensor

Relay Mirror

Test-Beam Generator

Directing the beam. A relay mirror reflects a high-powered laser beam from the ground station that generated it to a battle mirror more than a thousand miles away. A weaker test beam fired from a generator mounted on the mirror allows computers at the ground station to gauge and then compensate for atmospheric distortion *(box, below)*. A position sensor helps ensure that the two space-based mirrors are properly aligned, and communications antennae relay reports on hits and misses.

Eliminating Distortion

A laser beam's strength derives from the fact that all the light waves of which it is composed travel in phase with each other—peaks aligned with peaks and troughs with troughs, as represented by the aligned arrows in the drawings at right. But turbulence in the atmosphere can easily throw these waves out of phase, seriously affecting the beam's intensity and thus its ability to destroy targets. Compensating for this requires intricate computer analysis of waves and manipulation of a segmented mirror.

First, a test beam is sent from the relay mirror in space toward a mirror at the ground station *(top drawing)*. Distorted by the atmosphere *(blue band)*, the beam arrives out of phase *(staggered arrows)* and is reflected toward a computerized wave-front analyzer. Having determined the precise degree of the distortion, the computer then sends instructions to actuators mounted behind the ground-based mirror; the actuators deform the mirror by raising or lowering its thousands of individual segments. When the full-powered beam is fired *(bottom drawing)*, the deformed mirror distorts it exactly opposite to the way the atmosphere will; thus, as the waves pass through the atmosphere, they are redistorted back into phase.

Beam-Generating Station

Smart Rocks and Brilliant Pebbles

Although lasers would strive to knock out a high percentage of missiles in the very earliest stages after launch, they could not be expected to rebuff the attack entirely on their own. Other types of weapons would be needed to deal with the survivors and diversify the defense.

One of the most promising alternatives would be space-based interceptors—projectiles that would inflict damage by crashing into targets at tremendous speeds. Since the projectiles would carry no explosives, they would be relatively light and thus easier and cheaper to boost into orbit, so thousands could be incorporated into the defensive shield.

Intercepting projectiles could be deployed in several ways. Under one scheme, groups of interceptors would be housed in carrier satellites that would in turn be supervised by the battle-management network. Carrier satellites would be assigned their own sets of targets and instructed when and in what sequence to fire their interceptors, so that wasteful overlap could be kept to a minimum. Each carrier would then use its own sensors and onboard computers to aim its shots. To guarantee high kill rates, the interceptors would be equipped with sophisticated sensing and computing capabilities. In the example here of interceptors positioned to strike in the later stages of the boost phase, each so-called smart rock would be programmed to ferret out even missiles hidden by their own plumes (box, below right).

An alternative approach is to use smaller, more self-sufficient projectiles dubbed brilliant pebbles. Cheaper to launch and more highly computerized than smart rocks, pebble interceptors would be capable of identifying and intercepting targets without the assistance of carrier satellites or a battle-management network.

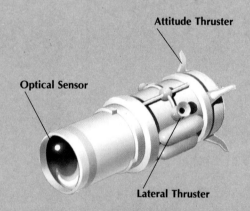

Optical Sensor

Attitude Thruster

Lateral Thruster

An impending collision. The final stage of the interceptor, separated from its boosters, closes on a missile at 25,000 miles per hour—so fast that impact alone, without an explosion, will destroy the target. An onboard microprocessor fires attitude thrusters to keep the projectile's optical sensor pointed directly at the missile, while the interceptor continues to lead its target. Based on the sensor's readings, the microprocessor will make last-second course corrections by firing lateral thrusters to push the interceptor up, down, or to left or right.

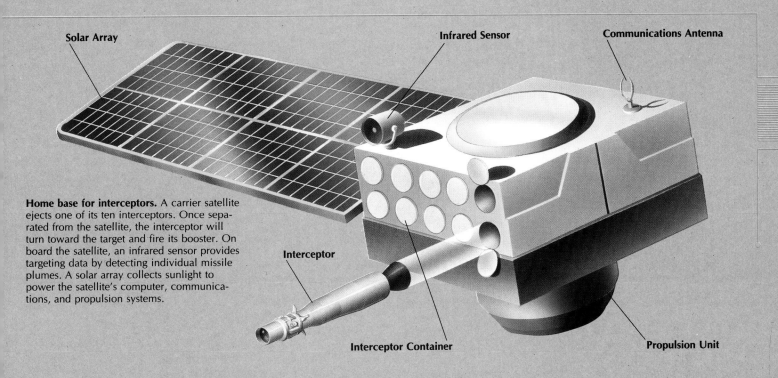

Solar Array

Infrared Sensor

Communications Antenna

Home base for interceptors. A carrier satellite ejects one of its ten interceptors. Once separated from the satellite, the interceptor will turn toward the target and fire its booster. On board the satellite, an infrared sensor provides targeting data by detecting individual missile plumes. A solar array collects sunlight to power the satellite's computer, communications, and propulsion systems.

Interceptor

Interceptor Container

Propulsion Unit

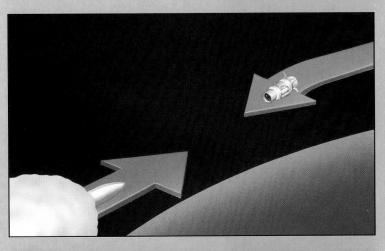

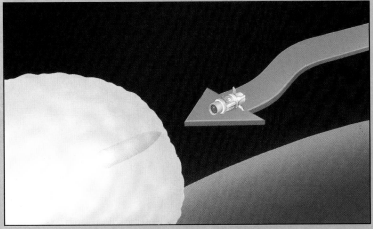

Staying on Course

Because it must make direct contact with its target, each interceptor must be able to make course corrections on its own. During early stages of the boost phase, the process, though involving complex navigational computations, is relatively straightforward: A microprocessor continuously monitors the interceptor's own trajectory and that of its target—denoted by the red arrows in the top drawing at left— then calculates where they will intersect and adjusts its flight path as necessary. But later, in the thin air of the upper atmosphere, a missile's plume will expand to envelop it *(bottom drawing)*, obscuring the missile's exact whereabouts from the interceptor's optical sensor. Any of several strategies to counter this problem could be programmed into the interceptor's microprocessor. In one proposed scheme, the microprocessor begins to compare current sensor readings with digitized plume images stored in its memory, which indicate the true positions of missiles within their plumes based on earlier tests. When the microprocessor finds a match, it can make the final adjustments to ensure a collision.

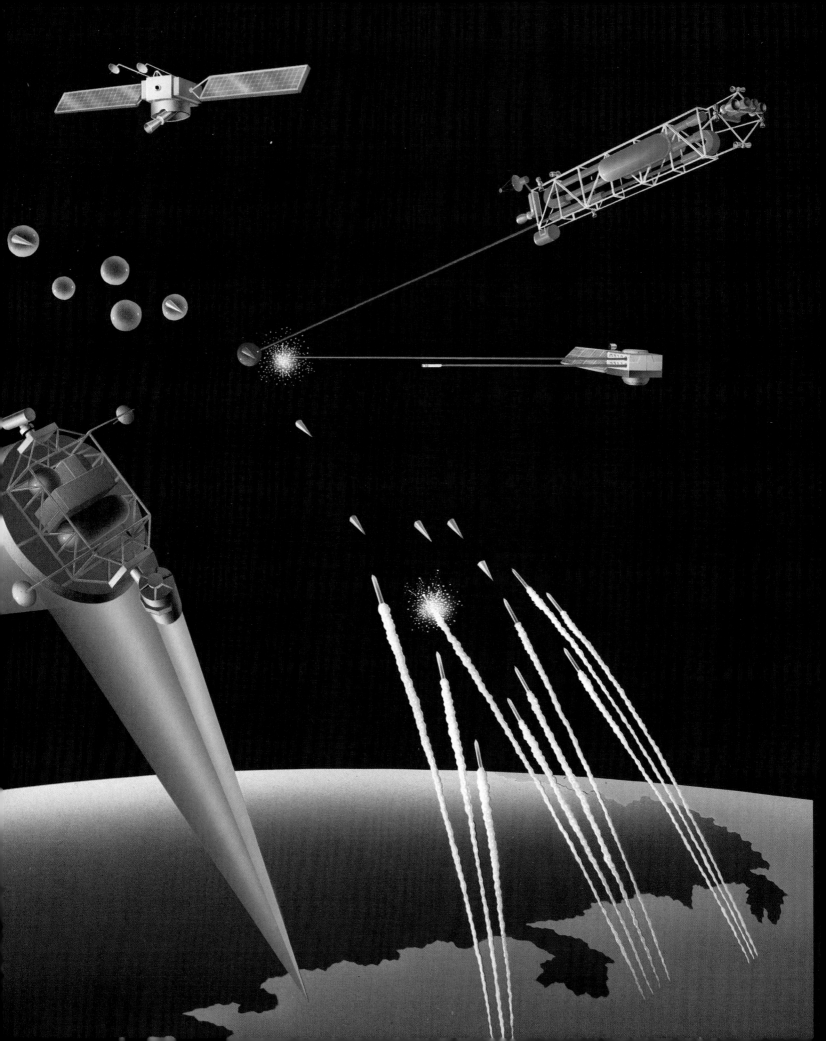

Bibliography

Books

Allen, Thomas B., *War Games.* New York: McGraw-Hill, 1987.

Andriole, Stephen J., ed., *High Technology Initiatives in C3I.* Washington: AFCEA International Press, 1986.

Bamford, James, *The Puzzle Palace.* New York: Penguin Books, 1983.

Barnaby, Frank, *The Automated Battlefield.* New York: Free Press, 1986.

Barnaby, Frank, ed., *Future War: Armed Conflict in the Next Decade.* New York: Facts on File, 1984.

Betts, Richard K., ed., *Cruise Missiles: Technology, Strategy, Politics.* Washington: Brookings Institution, 1981.

Blair, Bruce G., *Strategic Command and Control: Redefining the Nuclear Threat.* Washington: Brookings Institution, 1985.

Burleson, Clyde W., *The Jennifer Project.* Englewood Cliffs, N.J.: Prentice-Hall, 1977.

Burrows, William E., *Deep Black.* New York: Random House, 1986.

Canan, James W., *War in Space.* New York: Harper & Row, 1982.

Carter, Ashton B., John D. Steinbruner, and Charles A. Zraket, eds., *Managing Nuclear Operations.* Washington: Brookings Institution, 1987.

Computer Languages (Understanding Computers series). Alexandria, Va.: Time-Life Books, 1986.

Constant, James N., *Fundamentals of Strategic Weapons: Offense and Defense Systems.* The Hague: Martinus Nijhoff, 1981.

De Arcangelis, Mario, *Electronic Warfare.* Guildford, England: Blandford Press, 1985.

Doleman, Edgar C., Jr., and the editors of Boston Publishing Company, *Tools of War.* Boston: Boston Publishing, 1985.

Ford, Daniel, *The Button: The Pentagon's Strategic Command and Control System.* New York: Simon and Schuster, 1985.

Freidman, Richard S., et al., *Advanced Technology Warfare.* New York: Harmony Books, 1985.

Gerken, Louis, *ASW versus Submarine Technology Battle.* Chula Vista, Calif.: American Scientific, 1986.

Hirst, Mike, *Airborne Early Warning.* London: Osprey, 1983.

Jane's Weapon Systems 1985-86. London: Jane's Publishing Company, 1986.

Jenson, John R., *Introductory Digital Image Processing: A Remote Sensing Perspective.* Englewood Cliffs, N.J.: Prentice-Hall, 1986.

Karas, Thomas, *The New High Ground.* New York: Simon and Schuster, 1983.

Kennedy, William V., et. al., *Intelligence Warfare.* New York: Crescent Books, 1983.

Knightley, Phillip, *The Second Oldest Profession.* New York: W. W. Norton, 1986.

Krass, Allan S., *Verification: How Much Is Enough?* London and Philadelphia: Taylor & Francis, 1985.

Lee, R. G., *Introduction to Battlefield Weapons Systems.* London: Brassey's Defence Publishers, 1985.

Marchetti, Victor, and John D. Marks, *The CIA and the Cult of Intelligence.* New York: Alfred A. Knopf, 1974.

Paschall, Lee M., ed. *Air and Satellite Communications.* Washington: AFCEA International Press, 1985.

Richelson, Jeffrey, *The U.S. Intelligence Community.* Cambridge, Mass.: Ballinger Publishing, 1985.

Shaker, Steven M., and Alan R. Wise, *War without Men: Robots on the Future Battlefield.* McLean: Pergamon-Brassey's International Defense Publishers, 1988.

Stefanick, Tom, *Strategic Antisubmarine Warfare and Naval Strategy.* Lexington, Mass.: Lexington Books, 1987.

Walker, Bryce, and the Editors of Time-Life Books, *Fighting Jets* (The Epic of Flight series). Alexandria, Va.: Time-Life Books, 1983.

Werrell, Kenneth P., *The Evolution of the Cruise Missile.* Maxwell Air Force Base, Ala.: Air University Press, Sept. 1985.

Yost, Graham, *Spy-Tech.* New York: Facts on File, 1985.

Periodicals

Adam, John A., "Peacekeeping by Technical Means." *IEEE Spectrum,* July 1986.

Adam, John A., and Mark A. Fischetti, "SDI: The Grand Experiment." *IEEE Spectrum,* Sept. 1985.

Adam, John A., and Paul Wallich, "Mind-Boggling Complexity." *IEEE Spectrum,* Sept. 1985.

Amlie, Thomas, "How Radar Makes Our Smart Weapons Dumb." *Washington Post,* May 10, 1987.

Anderton, David A., "Tomorrow's Fighter: Updated or Outdated." *Aviation Week & Space Technology,* Aug. 10, 1987.

Antoniotti, Joseph C., "Precision-Guided Munitions: Semi-Active Laser versus Millimetre-Wave Guidance." *International Defense Review,* Sept. 1986.

Bell, D. Clifford, "Radar Countermeasures and Counter Countermeasures." *Military Technology,* May 1986.

Bethe, Hans A., et al., "Space-Based Ballistic-Missile Defense." *Scientific American,* Oct. 1984.

Beyers, Dan, "AMRAAM's Recent Test Success May Put Overdue Program on Track." *Defense News,* May 11, 1987.

Biddle, Wayne:
"How Much Bang for the Buck?" *Discover,* Sept. 1986.
"Star Wars: The Dream Diminished." *Discover,* July 1987.

Borning, Alan J., "Computer System Reliability and Nuclear War." *Communications of the ACM,* Feb. 1987.

Brand, David, "Star Wars' Hollow Promise." *Time,* Dec. 7, 1987.

Broad, William J., "Private Cameras in Space Stir U.S. Security Fears." *New York Times,* Aug. 25, 1987.

Brookner, Eli, "Phased Array Radars." *Scientific American,* Feb. 1985.

Browne, Malcolm W., "A Ship Killer Comes of Age." *Discover,* July 1982.

Cammarota, Richard S., "Defensive Watch." *Air Force Magazine,* Feb. 1985.

Canan, James W.:
"The ATF: Hot and Stealthy." *Air Force Magazine,* April 1987.
"The Emerging Lineup for SDI." *Air Force Magazine,* July 1987.

Carrington, Tim, "The Silent War: Undersea Arms Race Is Preoccupying Navies of U.S., Soviet Union." *Wall Street Journal,* June 24, 1987.

Christiansen, Donald, et al., "Star Wars, SDI: The Grand Experiment." *IEEE Spectrum,* Sept. 1985.

Clancy, George M., "New Breakthroughs in ASW Signal Processing." *Countermeasures,* Oct./Nov. 1975.

"Computer Sabotage: Programmed to Sneeze." *The Economist,* Nov. 28, 1987.

"Computerized Counterattack." *Science Digest,* June 1984.

Conniff, Richard, "Computer War." *Science Digest,* Jan. 1982.

Correll, John T., "Harvest and Seedtime in C3I." *Air Force Magazine,* June 1985.

Coyne, James P., "Electronics for the Shooting War." *Air Force Magazine,* June 1985.

Defense Science & Electronics, June 1985.

Doe, Charles, "Joint Tank Assault Weapon Nears Reality." *Navy Times,* Jan. 30, 1984.

Donnelly, Tom, "LTV, Martin FAAD Guns Lead Trials." *Defense News,* Oct. 12, 1987.

Dornheim, Michael A., "Rockwell to Test Space-Based Missile Interceptor for SDI." *Aviation Week & Space Technology,* Sept. 14, 1987.

Dugdale, Don, "Tapping the EW Potential of Unmanned Air Vehicles." *Defense Electronics,* Oct. 1986.

Dworetzky, Tom, "Run Silent, Run Deadly: The Race for Sneaky Subs." *Discover,* Dec. 1987.

"Electronic Warfare I/Electronic Warfare II." *Navy International,* Dec. 1985.

"Fast Track for C3I." *Air Force Magazine,* July 1984.

Foley, Theresa M., et al., "Strategic Defense Initiative: Blueprint for a Layered Defense." *Aviation Week & Space Technology,* Nov. 23, 1987.

Freedman, David, "Military Software's New Market." *High Technology Business,* Sept. 1987.

Gaskill, Jack D., ed. "Everything You Always Wanted to Know about Charge-Coupled Devices but Were Afraid to Ask." *Optical Engineering,* Aug. 1987.

Georg, Erich, "Advanced Infrared Homing Heads for Guided Missiles and Terminally Guided Munitions." *Military Technology,* June 1984.

Gilmartin, Trish, "SDI Organization Speeds Space-Based Interceptor Development." *Defense News,* Aug. 24, 1987.

Gooding, Judson, "Protector of the American Fleet." *New York Times Sunday Magazine,* Oct. 6, 1985.

Gordon, Don, "The JTIDS/PLRS Hybrid: A NATO Standard?" *Military Technology,* May 1987.

Gordon, Michael R., "House Panel Says MX Tests Indicate Serious Problems." *New York Times,* Aug. 24, 1987.

Greenwood, Ted, "Reconnaissance and Arms Control." *Scientific American,* Feb. 1973.

"Grumman Team to Develop Joint STARS." *Defense Electronics,* Nov. 1985.

Gwynne, Peter, "Sharper Eyes on the Sky." *High Technology,* Nov. 1986.

Hafemeister, David, "Advances in Verification Technology." *Bulletin of the Atomic Scientists,* Jan. 1985.

Hafemeister, David, Joseph J. Romm, and Kosta Tsipis, "The Verification of Compliance with Arms-Control Agreements." *Scientific American,* March 1985.

Heppenheimer, T. A.:
"The Real-Life Search for Red October." *Science Digest,* April 1986.
"Stopping Missiles with a Wham, Not a Zip." *High Technology,* March 1986.

Hessman, James D., "Anatomy of a Ship: The Aegis Program, and How It Grew." *Sea Power,* July 1986.

Hessman, James D., and Jack Hessman, "The Electronic Abacus: How Computers Have Changed the Navy." *Sea Power,* Aug. 1985.

Hornik, Richard, and Bruce van Voorst, "When Attackers Become Targets." *Time,* June 1, 1987.

Hubbell, John G., "You Are Under Attack!" *Reader's Digest,* April 1961.

Jackson, T. J., "Guided Bombs: Then and Now." *Military Technology,* June 1986.

Janesick, James R., "Charge-Coupled-Device Characterization, Modeling, and Application." *Optical Engineering,* Aug. 1987.

Janesick, James R., and Morley Blouke, "Sky on a Chip: The Fabulous CCD." *Sky & Telescope,* Sept. 1987.

Kelly, Donald W., and Mark C. Spear, "Lantirn: A Technical Report." *Defense Electronics,* Sept. 1986.

Kinnucan, Paul, "Building a Better Cockpit." *High Technology,* Jan. 1986.

Klass, Philip J.:
"Antiaircraft Warning Radar Plays Key Role in Survival." *Aviation Week & Space Technology,* April 7, 1986.
"Electronic Systems Emerge As Costliest Avionic Item." *Aviation Week & Space Technology,* April 7, 1986.
"Neutral Particle Beams Show Potential for Decoy Discrimination." *Aviation Week & Space Technology,* Dec. 8, 1986.
"New Guidance Technique Being Tested." *Aviation Week & Space Technology,* Feb. 25, 1974.
"Radar Countermeasures." *High Technology,* Aug. 1986.

Leopold, George, "Technology Enhances Target Recognition." *Defense News,* Oct. 5, 1987.

Lerner, Eric J.:
"Strategic C3: A Goal Unreached." *IEEE Spectrum,* Oct. 1982.
"Tactical C3: Survival under Duress." *IEEE Spectrum,* Oct. 1982.

Lerner, Michael A., "Are Tanks Here to Stay?" *Newsweek,* Sept. 27, 1982.

Lin, Herbert, "The Development of Software for Ballistic-Missile Defense." *Scientific American,* Dec. 1985.

MacKenzie, Donald, "Missile Accuracy—an Arms Control Opportunity." *Bulletin of the Atomic Scientists,* June/July 1986.

Magnuson, Ed, "D-Day in Grenada." *Time,* Nov. 7, 1983.

Marcus, Daniel J., George Leopold, and Tom Donnelly, "Command, Control and Communications." *Defense News,* Sept. 28, 1987.

"Mending America's Electronic Fences." *Defense Electronics,* Dec. 1985.

Moll, Kenneth L., "Understanding Command and Control." *Defense & Foreign Affairs Digest,* June 1978.

"Moon Stirs Scare of Missile Attack." *New York Times,* Dec. 8, 1960.

Moore, Richard L. "PAVE PAWS Protects the U.S. from Sea-Launched Missiles." *Defense Systems Review,* Aug. 1983.

Morrocco, John D.:
"Boeing Wins USAF Competition to Build Advanced Missile." *Aviation Week & Space Technology,* Dec. 15, 1986.
"Coming up Short in Software." *Air Force Magazine,* Feb. 1987.

Nelson, Greg, and David Redell, "The Star Wars Computer System." *Abacus,* Winter 1986.

Nordwall, Bruce D., "Advanced Cockpit Development Effort Signals Wide Industry Involvement." *Aviation Week & Space Technology,* April 20, 1987.

Oberg, James E., "Eyes on the Sky." *Air & Space,* April/May 1987.

Otten, David D., Elliot I. Bailis, and Jerry G. Klayman, "GEODSS: Heavenly Chronicler." *TRW/DSSG/QUEST,* Autumn 1980.

Patel, C. Kumar N., and Nicolaas Bloembergen, "Strategic Defense and Directed-Energy Weapons." *Scientific American,* Sept. 1987.

Preston, Antony, "Aegis—Shield of the Fleet." *Jane's Defense Weekly,* June 15, 1985.

Randolph, Anne, "USAF Upgrades Deep Space Coverage." *Aviation Week & Space Technology,* Feb. 28, 1983.

Rawles, Richard, "The Myth of Technology Transfer." *PC World,* May 1987.

Reed, Fred, "Let's Reform the Military Reformers." *Washington Post,* October 11, 1987.

Retelle, John P., Jr., and Mike Kaul, "The Pilot's Associate—Aerospace Application of Artificial Intelligence." *SIGNAL,* June 1986.

Richelson, Jeffrey, "The Keyhole Satellite Program." *Journal of Strategic Studies,* June 1984.

Richmond, Jeff, "In the Wings: Self-Repairing Jets." *High Technology,* July 1985.

Riordan, J. Timothy, "HgCdTe: Military High Flyer." *Photonics Spectra,* August 1984.

Russ, Robert D., "Spreading the Firepower, Extending the Battlefield." *Air Force Magazine,* April 1987.

Salamone, Salvatore:
"The Advanced Tactical Fighter: Smarter, Stronger, Nimbler." *High Technology,* Feb. 1987.
"Pentagon Sees Infrared." *High Technology,* May 1987.

Sanger, David E.:
"A Debate about 'Star Wars.' " *New York Times,* Oct. 13, 1985.
"U.S. Is Reviving Its Push to Build Fast Computers." *New York Times,* Aug. 13, 1987.

Schefter, Jim, "Stealthy Robot Planes." *Popular Science,* Oct. 1987.

Schultz, James, and David Russell, "New Staring Sensors." *Defense Electronics,* July 1984.

"Short Cuts to the Future." *Air Force Magazine,* Oct. 1987.

"The Silent War beneath the Waves." *U.S. News & World Report,* June 15, 1987.

Smith, Harry B., "Evolution of Radar Technology." *Signal,* July 1986.

Snyder, Samuel S., "Influence of U.S. Cryptologic Organizations on the Digital Computer Industry." *Journal of Systems and Software,* 1979.

Stein, Daniel L., "Would U.S. Communications Systems Work in a Nuclear War?" *Christian Science Monitor,* Nov. 17, 1982.

Teller, Edward, "Electromagnetic Pulses from Nuclear Explosions." *IEEE Spectrum,* Oct. 1982.

"Texas Instruments Boosts Reliability of Navy High-Speed Antiradiation Missile." *Aviation Week & Space Technology,* April 7, 1986.

Thompson, Stephen L., "The Big Picture." *Air & Space,* April/May 1987.

Troy, Charles T., "SDI Sets Its Sights on Large Detector Arrays." *Photonics Spectra,* Oct. 1986.

Tsipis, Kosta:
"The Accuracy of Strategic Missiles." *Scientific American,* July 1975.
"Arms Control Pacts Can Be Verified." *Discover,* April 1987.
"Cruise Missiles." *Scientific American,* Feb. 1977.

Ulsamer, Edgar, "C3I Keeps Climbing." *Air Force Magazine,* July 1985.

Vartabedian, Ralph, "AMRAAM Advanced Medium Range Air-to-Air Missile." *Los Angeles Times,* May 31, 1987.

"The Vast Potential of Tactical Technology." *Air Force Magazine,* April 1987.

"Virtual Cockpit's Panoramic Displays Afford Advanced Mission Capabilities." *Aviation Week & Space Technology,* Jan. 14, 1985.

Walker, Paul F.:
"Precision-Guided Weapons." *Scientific American,* Aug. 1981.
"Smart Weapons in Naval Warfare." *Scientific American,* May 1983.

Weizenbaum, Joseph, "Facing Reality: Computer Scientists Aid War Efforts." *Technology Review,* Jan. 1987.

Worden, Simon, "Lasers for Defense." *NATO's Sixteen Nations,* June 1987.

Other Publications

Antoniotti, Joseph C., "Copperhead Artillery's Tank Killer." Unpublished paper.

Baker, William R., and Roger W. Clem, *Terrain Contour Matching (TERCOM) Primer.* Technical Report ASD-TR-77-61: Wright-Patterson Air Force Base, Ohio: Aug. 1977.

Conrow, E. H., G. K. Smith, and A. A. Barbour, "The Joint Cruise Missiles Project: An Acquisition History—Appendixes." *A Rand Note,* N-1989-JCMPO: Aug. 1982.

DARPA, *Strategic Computing, Third Annual Report.* Feb. 1987.

Davis, Paul K., Steven C. Banks, and James P. Kahan, "A New Methodology for Modeling National Command Level Decisionmaking in War Games and Simulations." Santa Monica, Calif.: RAND Corp., July 1986.

Donohue, G., B. Bennett, and J. Hertzog, "The RAND Military Operations Simulation Facility: An Overview." *A RAND Note.* Santa Monica, Calif.: RAND Corp., April 1984.

Eastport Study Group, "A Report to the Director, Strategic Defense Initiative Organization." Dec. 1985.

EMI/RFI Shielding Systems and Accessories. Lindgren RF Enclosures.

Fletcher, James C., et al., "Report of the Study on Eliminating the Threat Posed by Nuclear Ballistics Missiles." Feb. 1984.

Furness, Thomas A., III, "The Super Cockpit and Its Human Factors Challenges." *Proceedings of the Human Factors Society, 30th Annual Meeting,* 1986.

General Dynamics F-16C: An Evolutionary Change for Increased Combat Capability. Fort Worth, Texas: General Dynamics.

Golden, Joe P., "Terrain Contour Matching (TERCOM): A Cruise Missile Guidance Aid." *Proceedings of the Society of Photo-Optical Instrumentation Engineers, Volume 238, Image Processing for Missile Guidance.* Bellingham, Wash.

Goldsmith, Martin, "Applying the National Training Center Experience—Incidence of Ground-to-Ground Fratricide." *A RAND Note.* Santa Monica, Calif.: RAND Corp., Feb. 1986.

"Hardened Target Weapons." *Fact Sheet:* United States Air Force Office of Public Affairs, Jan. 1987.

Hudson, C. K., "Aquila—The Army Remotely Piloted Vehicle." Paper presented at the 12th Annual Meeting of the Association for Unmanned Vehicle Systems, Anaheim, Calif., July 15-17, 1985.

"The JANUS Interactive Tactical Simulation." Santa Monica, Calif.: RAND Corp., July 9, 1987.

Klahr, Philip, and John W. Ellis, Jr., et al., "TWIRL: Tactical Warfare in the ROSS Language." Santa Monica, Calif.: RAND Corp., Oct. 1984.

Marquet, Louis C., "The Strategic Defense Initiative, Progress and Prospects." Paper delivered Jan. 30, 1987.

"The National Training Center (NTC) Replay System." Santa Monica, Calif.: RAND Corp., July 23, 1987.

"PAVEWAY III." *Fact Sheet:* United States Air Force Office of Public Affairs, April 1987.
"The Pilot's Associate." United States Air Force Fact Sheet PAM #86-039. Wright-Patterson Air Force Base: 1986.
Sims, Richard E., "A HEMP Hardened Fiber Optic Repeater Terminal." Paper presented to the First International Military and Government Fiber Optics and Communications

Exposition, March 19, 1987.
"Strategic Defense Initiative (SDI) Models." Santa Monica, Calif.: RAND Corp., July 23, 1987.
"The TAC BRAWLER." Santa Monica, Calif.: RAND Corp., July 23, 1987.
Westinghouse Electric Corporation, "Electronic Warfare." Booklet, Sept. 1976.

Acknowledgments

The index for this book was prepared by Mel Ingber. The editors also wish to thank: **In Canada:** Ontario—David L. Parnas, Queen's University; **In France:** Paris—Hélène Mounier, Editions Lariviere; **In Germany**—Bremen: Kurt Stolz, MBB, Marine and Special Products Division; **In Italy**—Genoa: Michele Nunes; La Spezia: Berto Caporali, Selenia S.P.A.; Rome: Anna-Maria Capparelli and Marco Caporali, Agusta S.P.A.; Contessa Maria Fede Caproni; **In the United States:** California—Palo Alto: Gary Chapman and Katy Elliot, Computer Professionals for Social Responsibility; George Nercanchi, Defense Electronics; Redondo Beach: Tim Dolan and Edie Cartwright, TRW Electronics & Defense Sector; San Jose: Andreas Kyriakou, Rolm Mil-Spec Computers; Santa Monica: Pat Allen, Bart Bennett, William Giarla, Gail Halverson, Ted Harshberger, Jon B. Hertzog, Jim Hodges, Rich Mills, and Tim Webb, RAND Corporation; Stanford: Harvey Lynch; Colorado—Colorado Springs: Nicholas L. Johnson, Teledyne Brown Engineering; Connecticut—Westport: Judith Paris Roth, Optical Information Systems Journal; District of Columbia—Dr. Thomas Amlie, SAF/ACS, The Pentagon; Tom Blau, Institute of Foreign Policy Analysis; Maj. Alan Freitag, Col. Ralph Gajewski, Neil Griff, Comdr. David Newton, Maj. William O'Connell, Lt. Col. James Price, and Col. Raymond Ross, Strategic Defense Initiative Organization; Jane Gruenebaum; Susan K. Hopkins and Robert E. Holsapple, Cruise Missiles Project Office; Dick King, Westinghouse Electric; John Shore, Entropic Processing, Inc.; Josephine Anne Stein, Sandy Thomas, and Anna C. Urband, Department of the Navy; Douglas Waller; Florida—Eglin Air Force Base: Senior Airman Richard J. Bartleson, Jr., Office of Public Affairs, Armament Division; Maryland—Baltimore: Robert F. Jolin, Westinghouse Defense and Electronics Center; Rockville: Stan Fishkind, ORI, Inc.; Silver Spring: Del Malkie; Massachusetts—Amherst: Allan Krass, Hampshire College; Buzzard's Bay: Capt. Thomas D. Parker, Cape Cod AFS; Cambridge: Peter Clausen, Union of Concerned Scientists; Hanscom Air Force Base: Kevin Gilmartin; Marlboro: Geoffrey E. Smith, Raytheon Company; New Jersey—Morristown: Bonnie Hawkins and Marion Vanhorn Musto, Computer Sciences Corporation; New Mexico—Kirtland AFB: Juventino R. Garcia, Air Force Weapons Laboratory; Los Alamos: Barbara Mulkin and Bill Jack Rodgers, Los Alamos National Laboratory; White Sands Missile Range: Monte Marlin and Lt. Col. Edgar Williams, U.S. Army; Pennsylvania—State College: Don E. Gordon, HRB-Singer, Inc.; Virginia—Alexandria: Tom Mattson, Center for Naval Analysis; Kenneth L. Moll, Strategy Corporation; SDI Library, Institute for Defense Analysis; Arlington: John Bosma and Carolyn Meinel; Maj. David Nicholson, DARPA; Adm. Albert Gallotta; Crystal City: Comdr. Kate Paige, Naval Sea Systems Command; Falls Church: Donald F. Hemenway, Jr., American Society for Photogrammetry and Remote Sensing; Walter N. Lang; Donald C. Latham, Computer Sciences Corporation; Fairfax: James Coyne and Robert K. Ackerman, Signal Magazine; James Westwood, Military Science & Defense Analytics; Fairfax Station: Steven M. Shaker; McLean: Leroy B. Van Brunt, EW Engineering, Inc.; Woodbridge: Edward Walsh, Defense Technology Viewpoint.

Picture Credits

Index

Numerals in italics indicate an illustration of the subject mentioned.

TIME-LIFE BOOKS

EDITOR-IN-CHIEF: Thomas H. Flaherty

Director of Editorial Resources: Elise D. Ritter-Clough
Executive Art Director: Ellen Robling
Director of Photography and Research:
John Conrad Weiser
Editorial Board: Dale M. Brown, Janet Cave,
Roberta Conlan, Robert Doyle, Laura Foreman,
Jim Hicks, Rita Thievon Mullin, Henry Woodhead
Assistant Director of Editorial Resources:
Norma E. Shaw

PRESIDENT: John D. Hall

Vice President and Director of Marketing:
Nancy K. Jones
Editorial Director: Lee Hassig
Director of Production Services: Robert N. Carr
Production Manager: Marlene Zack
Director of Technology: Eileen Bradley
Supervisor of Quality Control: James King

Editorial Operations
Production: Celia Beattie
Library: Louise D. Forstall
Computer Composition: Deborah G. Tait (Manager),
Monika D. Thayer, Janet Barnes Syring, Lillian Daniels
Interactive Media Specialist: Patti H. Cass

Time-Life Books is a division of
Time Life Incorporated

PRESIDENT AND CEO: John M. Fahey, Jr.

Correspondents: Maria Vincenza Aloisi (Paris); Ann
Natanson (Rome); Dick Berry (Tokyo). Valuable assistance was also provided by: Angie Lemmer (Bonn);
Elizabeth Brown and Christina Lieberman (New York).

UNDERSTANDING COMPUTERS

SERIES DIRECTOR: Lee Hassig
Series Administrator: Loretta Britten

Editorial Staff for *The Military Frontier*
Designer: Christopher M. Register
Associate Editors: Jeremy Ross, Kristin Baker (pictures),
Allan Fallow (text)
Researchers: Steven Feldman, Flora Garcia, Gretchen
Luchsinger, Barbara J. Scruggs
Writer: Robert M. S. Somerville
Assistant Designers: Paul M. Graboff, Tessa
Tilden-Smith
Copy Coordinator: Elizabeth Graham
Picture Coordinators: Renée DeSandies, Linda Yates
Editorial Assistant: Susan L. Finken

Special Contributors: Wayne Biddle, James W. Canan,
Elizabeth Carpenter, Philip J. Klass, Robert Lee
Permenter, Mitchell M. Waldrop (text); Holly Knox,
Lucinda Moore, Julie Ann Trudeau (research).

CONSULTANTS

THOMAS B. ALLEN is the author of *War Games: The
Secret World of the Creators, Players, and Policy Makers
Rehearsing World War III Today.* He is the coauthor (with
Norman Polmar) of *Rickover: Controversy and Genius.*

JOSEPH C. ANTONIOTTI is director of studies and analyses of Sverdrup Technology, an engineering consultant
firm in Tullahoma, Tennessee. He has written numerous
articles about precision-guided munitions.

ELI BROOKNER is a radar expert and consulting scientist
for the Raytheon Company in Wayland, Massachusetts.
He has written about the subject of phased-array radar for
Scientific American and is the principal author of *Radar
Technology.*

WILLIAM E. BURROWS, the director of New York University's Science and Environmental Reporting Program,
is the author of *Deep Black: Space Espionage and National Security.*

THOMAS A. FURNESS III is chief of Visual Display Systems at Wright-Patterson Air Force Base near Dayton,
Ohio. He pioneered the concept of the Supercockpit.

JOHN HILL is senior military analyst for the Orkand
Corporation, a consulting firm that specializes in national
security analysis.

PHILIP J. KLASS is a contributing avionics editor for *Aviation Week & Space Technology.* He served as senior
avionics editor for that magazine for thirty-five years.

MARVIN LEIBSTONE is the North American editor-in-chief for the Monch Publishing Group, publishers of
Military Technology, Naval Forces, and *NATO's 16 Nations* magazines.

JOHN PIKE is associate director for space policy at the
Federation of American Scientists in Washington, D.C.

FRED REED, a Marine veteran of Vietnam, has covered
military affairs for numerous publications. He writes the
syndicated column "Soldiering" and is the Washington
editor for *Harper's* magazine.

WILLIAM A. SKILLMAN is a consulting engineer with the
Westinghouse Defense and Electronics Center in Baltimore, Maryland.

TOM STEFANICK, who holds a fellowship in Science,
Arms Control, and National Security from the American
Academy for the Advancement of Science, is a member
of the House Armed Services Committee. He is the author
of *Strategic Anti-Submarine Warfare and Naval Strategy.*

Commander L. T. TERRELL, U.S.N. (Ret.), served as the
combat-systems officer aboard the U.S.S. *Yorktown,* the
second Aegis cruiser commissioned by the U.S. Navy.

PETER ZIMMERMAN is director of the Project on SDI
Technology and Policy at the Carnegie Endowment for
International Peace in Washington, D.C.

Library of Congress Cataloging in Publication Data

The Military frontier / by the editors of Time-Life Books.
    p.   cm.—(Understanding computers)
    Includes bibliographical references and index.
    ISBN 0-8094-7610-X—ISBN 0-8094-7611-8 (lib. bdg.)
    1. Military art and science—Data processing.
2. Military art and science—Automation.
3. Artificial intelligence—Military applications.
I. Series.
UG478.M55 1991
623'.0285'63—dc20   91-13766
                                                            CIP

For information on and a full description of any of the Time-Life Books series listed, please call 1-800-621-7026 or write:
Reader Information
Time-Life Customer Service
P.O. Box C-32068
Richmond, Virginia 23261-2068

REVISIONS STAFF

EDITOR: Lee Hassig

Designer: Christopher M. Register

Writer: Esther Ferington
Assistant Designer: Bill McKenney
Copy Coordinator: Donna Carey
Picture Coordinator: Leanne Miller

Consultants:
ANTON COLIJN is a professor of computer science
at the University of Calgary.

FRED REED (see CONSULTANTS).